Collins revision guides

D1324317

Paul Metcalf

Series Editor: Jayne de Courcy

Instant Revision

GCSE Maths

Contents

Published by HarperCollins*Publishers* Ltd
77–85 Fulham Palace Road
London W6 8JB

www.**Collins**Education.com
On-line support for schools and colleges

First published 2001
This new format edition published 2004

ISBN 0 00 717259 1

British Library Cataloguing in Publication Data
A catalogue record for this publication is available from the British Library

Edited by Joan Miller
Production by Katie Butler
Design by Gecko Limited
Printed and bound by Printing Express, Hong Kong

Acknowledgements

Illustrations
Gecko Ltd

Every effort has been made to contact the holders of copyright material, but if any have been inadvertently overlooked, the Publishers will be pleased to make the necessary arrangements at the first opportunity.

50p 3/6

Get the most out of your Instant Revision pocket book

1 **Maximise your revision time.** You can carry this book around with you anywhere. This means you can spend any spare moments dipping into it.

2 **Learn and remember what you need to know.** This book contains all the really important things you need to know for your exam. All the information is set out clearly and concisely, making it easy for you to revise.

3 **Find out what you don't know.** The *Check yourself* questions and *Score chart* help you see quickly and easily the topics you're good at and those you're not so good at.

What's in this book?

1 The facts – just what you need to know

- There are sections covering all the key skills and concepts tested in the Intermediate and Higher Tier GCSE Maths exams by all Boards.
- The Higher Tier sections cover A* and A grade topics. These pages are clearly marked and you only need to use these, as well as the other pages, if you will be sitting the Higher Tier papers.
- All the sections contain worked examples with clear commentaries on them.

2 *Check yourself* questions – find out how much you know and boost your grade

- Each *Check yourself* is linked to one or more facts page. The numbers after the topic heading in the *Check yourself* tell you which facts page the *Check yourself* is linked to.
- The questions are based on actual exam questions and will show you what you do and don't know.
- The reverse side of each *Check yourself* gives you the answers **plus** tutorial help and guidance to boost your exam grade.
- There are marks for each question. The total number of marks for each *Check yourself* is always 20. When you check your answers, fill in the score box alongside each answer with the number of marks you feel you scored: award yourself a proportion of the marks for partially correct answers.

3 The *Score* chart – an instant picture of your strengths and weaknesses

- There is one *Score chart* for Intermediate Tier *Check yourself* questions and one for the additional Higher Tier questions.
- *Score chart*(1) lists all the *Check yourself* pages.
- As you complete each *Check yourself*, record your marks on the *Score chart*. This will show you instantly which areas you need to spend more time revising.
- *Score chart*(2) is a graph which lets you plot your marks against GCSE grades. This will give you a rough idea of how you are doing in each area.
- If your grades on the Higher Tier *Check yourself* questions are higher than your grades on the Intermediate questions, then you need to spend more time revising the Intermediate topics.

Use this Instant Revision pocket book on your own – or revise with a friend or relative. See who can get the highest score!

Definition of numbers

Sets of numbers can be described in different ways:

natural numbers	1, 2, 3, 4, 5, ...
positive integers	$^{+}1, {}^{+}2, {}^{+}3, {}^{+}4, {}^{+}5, \ldots$
negative integers	$^{-}1, {}^{-}2, {}^{-}3, {}^{-}4, {}^{-}5, \ldots$
square numbers	1, 4, 9, 16, 25, 36, ...
triangle numbers	1, 3, 6, 10, 15, 21, ...

Multiples

The **multiples** of a number are the products of the multiplication tables.

e.g. Multiples of 3 are 3, 6, 9, 12, 15, 18, 21, 24, ...
Multiples of 4 are 4, 8, 12, 16, 20, 24, 28, 32, ...

The **least common multiple** (LCM) is the least multiple which is common to all of the given numbers.

e.g. Common multiples of 3 and 4 are 12, 24, 36, ...
The least common multiple is 12.

Factors

The **factors** of a number are the natural numbers which divide *exactly* into that number (i.e. without a remainder).

e.g. Factors of 8 are 1, 2, 4 and 8.
Factors of 12 are 1, 2, 3, 4, 6 and 12.

The **highest common factor** (HCF) is the highest factor which is common to all of the given numbers.

e.g. Common factors of 8 and 12 are 1, 2 and 4.
The highest common factor is 4.

Prime numbers

A **prime number** is a natural number with exactly two factors (i.e. 1 and itself).

The following numbers have exactly two factors so are prime numbers.
2, 3, 5, 7, 11, 13, 17, 19, 23, 29, 31, ...

Prime factors

A **prime factor** is a factor which is also prime.
All natural numbers can be written as a product of prime factors.

e.g. 21 can be written as 3×7 where 3 and 7 are prime factors.
60 can be written as $2 \times 2 \times 3 \times 5$ where 2, 3 and 5 are prime factors.

The prime factors of a number can be found by successively rewriting the number as a product of prime numbers in increasing order (i.e. 2, 3, 5, 7, 11, 13, 17, ... etc.).

e.g. $84 = 2 \times 42$ Writing 84 as 2×42.
$= 2 \times 2 \times 21$ Writing 42 as 2×21.
$= 2 \times 2 \times 3 \times 7$ Writing 21 as 3×7.

Squares

Square numbers are produced when numbers are multiplied by themselves.

e.g. The square of 8 is $8 \times 8 = 64$ and 64 is a square number.

Cubes

Cube numbers are produced when numbers are multiplied by themselves then multiplied by themselves again.

e.g. The cube of 5 is $5 \times 5 \times 5 = 125$ and 125 is a cube number.

Square roots

The **square root** of a number such as 36 is the number which when squared equals 36 i.e. 6 or $^-6$ (because $6 \times 6 = 36$ and $^-6 \times {}^-6 = 36$).

The sign $\sqrt[2]{\ }$ or $\sqrt{\ }$ is used to denote the square root. $\sqrt[2]{36}$ or $\sqrt{36} = \pm 6$

Cube roots

The **cube root** of a number such as 27 is the number which when cubed equals 27 i.e. 3 (because $3 \times 3 \times 3 = 27$).

The sign $\sqrt[3]{\ }$ is used to denote the cube root. $\sqrt[3]{27} = 3$

Reciprocals

The **reciprocal** of any number can be found by converting the number to a fraction and turning the fraction upside-down. The reciprocal of $\frac{2}{3}$ is $\frac{3}{2}$ and the reciprocal of 10 is $\frac{1}{10}$ (as $10 = \frac{10}{1}$).

Directed numbers

A **directed number** is one which has a + or – sign attached to it.

When adding or subtracting directed numbers, remember that signs written next to each other can be replaced by a single sign as follows.

+ + is the same as +	+ – is the same as –
– + is the same as –	– – is the same as +

e.g. $(^-1) + (^-2) = {}^-1 - 2 = {}^-3$ + – is the same as –

$(^+2) - (^-3) = {}^+2 + 3 = {}^+5$ – – is the same as +

To multiply or divide directed numbers, include the sign according to the following rules.

- if the signs are the same, the answer is positive
- if the signs are opposite, the answer is negative

or

+ × + = +	– × – = +	+ ÷ + = +	– ÷ – = +
+ × – = –	– × + = –	+ ÷ – = –	– ÷ + = –

e.g.

$(^-8) \times (^+2) = {}^-16$ – × + = –

$(^+12) \div (^-4) = {}^-3$ + ÷ – = –

$(^-2) \div (^-5) = {}^+\frac{2}{5}$ – ÷ – = +

$(^-5)^2 = {}^+25$ $(^-5)^2 = (^-5) \times (^-5)$ and – × – = +

Positive, negative and zero indices

When multiplying a number by itself you can use the following shorthand.

$7 \times 7 = 7^2$ You say '7 to the power 2' (or '7 squared').

$7 \times 7 \times 7 = 7^3$ You say '7 to the power 3' (or '7 cubed').

Multiplying indices

You can multiply numbers with indices as shown.

$$7^4 \times 7^6 = (7 \times 7 \times 7 \times 7) \times (7 \times 7 \times 7 \times 7 \times 7 \times 7)$$
$$= 7 \times 7 \times 7 \times 7 \times 7 \times 7 \times 7 \times 7 \times 7 \times 7$$
$$= 7^{10}$$

A quicker way to multiply two numbers with indices *when their bases are the same* is to add their powers.

$7^4 \times 7^6 = 7^{4+6} = 7^{10}$ and in general

$\boldsymbol{a^m \times a^n = a^{m+n}}$

Dividing indices

You can divide numbers with indices as shown.

$$5^6 \div 5^4 = \frac{5 \times 5 \times 5 \times 5 \times 5 \times 5}{5 \times 5 \times 5 \times 5}$$
$$= 5 \times 5$$
$$= 5^2$$

A quicker way to divide two numbers with indices *when their bases are the same* is to subtract their powers:

$5^6 \div 5^4 = 5^{6-4} = 5^2$ and in general

$\boldsymbol{a^m \div a^n = a^{m-n}}$

Negative powers

You know that $8^4 \div 8^6 = 8^{4-6} = 8^{-2}$

and $8^4 \div 8^6 = \dfrac{8 \times 8 \times 8 \times 8}{8 \times 8 \times 8 \times 8 \times 8 \times 8} = \dfrac{1}{8^2}$

so $8^{-2} = \dfrac{1}{8^2}$ and in general

$$\boldsymbol{a^{-m} = \frac{1}{a^m}}$$

e.g. $3^{-2} = \dfrac{1}{3^2} = \dfrac{1}{9}$ and $2^{-4} = \dfrac{1}{2^4} = \dfrac{1}{16}$ etc.

Zero powers

You know that $5^2 \div 5^2 = 5^{2-2} = 5^0$

and $5^2 \div 5^2 = \dfrac{5 \times 5}{5 \times 5} = 1$

so $5^0 = 1$ and in general

$\boldsymbol{a^0 = 1}$

(i.e. any number raised to the power zero is equal to 1)

e.g. $5^0 = 1$, $100^0 = 1$, $2.5^0 = 1$ and $(^-6)^0 = 1$ etc.

Significant figures

Any number can be rounded off to a given number of significant figures (written s.f.) using the following rules.

- Count along to the number of significant figures required.
- Look at the next significant digit.

  If its value is smaller than 5, leave the 'significant' digits as they are.

  If its value is 5 or greater, add 1 to the last of the 'significant' digits.
- Restore the number to its correct size by filling with zeros if necessary.

e.g. Round 547.36 to 4, 3, 2, 1 significant figures.

547.36 = 547.4 (4 s.f.)
547.36 = 547 (3 s.f.)
547.36 = 550 (2 s.f.)*
547.36 = 500 (1 s.f.)*

NB You need to pad the numbers marked with an asterisk (*) with zeros in order to restore the numbers to their correct size.

Decimal places

Any number can be rounded to a given number of decimal places (written d.p.) using the following rules.

- Count along to the number of the decimal places required.
- Look at the digit in the next decimal place.

  If its value is smaller than 5, leave the preceding digits (the digits before it) as they are.

  If its value is 5 or greater, add 1 to the preceding digit.
- Restore the number by replacing any numbers to the left of the decimal point.

e.g. Round 19.3461 to 4, 3, 2, 1 decimal places.

19.3461 = 19.3461 (4 d.p.)
19.3461 = 19.346 (3 d.p.)
19.3461 = 19.35 (2 d.p.)
19.3461 = 19.3 (1 d.p.)

NB The numbers to the left of the decimal point are not affected by this rounding process as you are only concerned with decimal places.

Without a calculator

Practise working without a calculator wherever possible. The following examples illustrate some common non-calculator methods.

Multiplying

e.g. Calculate 167×53

$$167 \times 53 = 167 \times (50 + 3) = 167 \times 50 + 167 \times 3 = 8350 + 501 = 8851$$

It is more usual to set this multiplication out like this.

167	
× 53	
8350	Multiplying by 50.
+ 501	Multiplying by 3.
8851	Adding.

Multiplying decimals

To multiply two decimals without using a calculator:

- ignore the decimal points and multiply the numbers
- add the number of digits after the decimal point in the numbers in the question
- position the decimal point so that the number of digits after the decimal point in the answer is the same as the total number of decimal places in the question.

e.g. Calculate 1.67×5.3

$167 \times 53 = 8851$ Ignoring the decimal points and multiplying the numbers.

The number of digits after the decimal point in the numbers $= 2 + 1 = 3$.

$1.67 \times 5.3 = 8.851$ Replacing the decimal point so that the number of digits after the decimal point in the answer is 3.

It is helpful to check that the answer is approximately correct i.e. 1.67×5.3 is approximately $2 \times 5 = 10$ so the answer of 8.851 looks correct.

Dividing

To divide by a two-digit number, proceed in exactly the same way as for any other division.

e.g. Calculate $513 \div 19$

```
    27
19)513
   38↓
   133
   133
     0
```

Dividing decimals

You can use the idea of equivalent fractions to divide decimals.

e.g. Work out 0.003 08 ÷ 0.000 14

$$0.003\,08 \div 0.000\,14 = \frac{0.003\,08}{0.000\,14}$$

$$= \frac{308}{14}$$ Multiplying top and bottom by 100 000 to obtain an equivalent fraction.

Now divide 308 ÷ 14

$$\begin{array}{r} 22 \\ 14\overline{)308} \\ \underline{28}\downarrow \\ 28 \\ \underline{28} \\ \underline{0} \end{array}$$

So 0.003 08 ÷ 0.000 14 = 22

Estimation and approximation

It is useful to check your work by approximating your answers to make sure that they are reasonable. Estimation and approximation questions are popular questions on the examination specification or syllabus. You will usually be required to give an estimation by rounding numbers to 1 (or 2) significant figures.

e.g. Estimate the value of $\frac{6.98 \times (10.16)^2}{9.992 \times \sqrt{50}}$

Rounding the figures to 1 significant figure and approximating $\sqrt{50}$ as 7:

$$\frac{6.98 \times (10.16)^2}{9.992 \times \sqrt{50}} \approx \frac{7 \times 10^2}{10 \times 7} = \frac{700}{70}$$

$$= 10$$

A calculator gives an answer of 10.197 777 so that the answer is quite a good approximation.

Imperial/metric units

In number work, it is common to be asked to convert between imperial and metric units. In particular, the following conversions may be tested in the examination.

Imperial	Metric
1 inch	2.5 centimetres
1 foot	30 centimetres
5 miles	8 kilometres
1 litre	1.75 pints
1 gallon	4.5 litres
2.2 pounds	1 kilogram

Fractions

The top part of a fraction is called the **numerator** and the bottom part is called the **denominator**.

Equivalent fractions

Equivalent fractions are fractions which are equal in value to each other. The following fractions are all equivalent to $\frac{1}{2}$.

$\frac{1}{2} = \frac{2}{4} = \frac{3}{6} = \frac{5}{10} = \ldots$

Equivalent fractions can be found by multiplying or dividing the numerator and denominator by the same number.

$\frac{3}{4} \xrightarrow{\times 10} \frac{30}{40}$ $\quad \frac{4}{7} \xrightarrow{\times 3} \frac{12}{21}$ $\quad \frac{160}{200} \xrightarrow{\div 2} \frac{80}{100}$ $\quad \frac{9}{12} \xrightarrow{\div 3} \frac{3}{4}$

i.e. $\frac{3}{4} = \frac{30}{40}$ (×10), $\frac{4}{7} = \frac{12}{21}$ (×3), $\frac{160}{200} = \frac{80}{100}$ (÷2), $\frac{9}{12} = \frac{3}{4}$ (÷3)

One number as a fraction of another

To find one number as a fraction of another, write the numbers in the form of a fraction.

e.g. Write 4 mm as a fraction of 8 cm.

First ensure that the units are the same.
Remember 8 cm = 80 mm.

4 mm as a fraction of 80 mm $= \frac{4}{80} = \frac{1}{20}$

So 4 mm is $\frac{1}{20}$ of 8 cm.

Addition and subtraction

Before adding (or subtracting) fractions, ensure that they have the same denominator.

e.g. $\frac{7}{8} - \frac{1}{5}$

$= \frac{35}{40} - \frac{8}{40}$ Writing both fractions with a denominator of 40, $\frac{7}{8} = \frac{35}{40}$ (×5) and $\frac{1}{5} = \frac{8}{40}$ (×8).

$= \frac{27}{40}$

To find the **common denominator** of two numbers, find their **least common multiple** or LCM.

The LCM of 8 and 5 is 40 (see card 1).

Multiplication of fractions

To multiply fractions, multiply the numerators and multiply the denominators.

e.g. $\frac{4}{7} \times \frac{2}{11}$

$= \frac{4 \times 2}{7 \times 11}$ Multiplying the numerators and multiplying the denominators.

$= \frac{8}{77}$

e.g. $1\frac{1}{5} \times 6\frac{2}{3}$

$= \frac{6}{5} \times \frac{20}{3}$ Converting to top-heavy fractions.

$= \frac{6 \times 20}{5 \times 3}$ Multiplying the numerators and multiplying the denominators.

$= \frac{120}{15} = 8$

NB Remember to cancel the fractions where possible.

Division of fractions

To divide one fraction by another, multiply the first fraction by the reciprocal of the second fraction.

e.g. $\frac{3}{7} \div \frac{1}{7}$

$= {}_1\frac{3}{\cancel{7}} \times \frac{\cancel{7}^1}{1}$ Cancelling fractions.

$= 3$ As $\frac{3}{1} = 3$.

e.g. $4\frac{4}{5} \div 1\frac{1}{15}$

$= \frac{24}{5} \div \frac{16}{15}$ Converting to top-heavy fractions.

$= \frac{{}^3\cancel{24}}{{}_1\cancel{5}} \times \frac{\cancel{15}^3}{\cancel{16}_2}$ Cancelling fractions.

$= \frac{3 \times 3}{1 \times 2}$

$= \frac{9}{2}$

$= 4\frac{1}{2}$ Rewriting as a mixed number.

Fractions to decimals

A fraction can be changed to a decimal by carrying out the division.

e.g. Change $\frac{3}{8}$ to a decimal.

$\frac{3}{8} = 3 \div 8 = 0.375$

e.g. Change $\frac{4}{15}$ to a decimal.

$\frac{4}{15} = 4 \div 15 = 0.266\,666\,6\ldots$

The decimal 0.266 666 6... carries on infinitely and is called a **recurring** decimal.

You can write the recurring decimal 0.266 666 6... as $0.2\dot{6}$. The dot over the 6 means that the number carries on infinitely.

If a group of numbers carries on infinitely, two dots can be used to show the repeating numbers.

$0.\dot{3}\dot{5} = 0.353\,535\,35\ldots$

$6.4\dot{1}\dot{7} = 6.417\,171\,717\ldots$

$3.\dot{2}0\dot{1} = 3.201\,201\,201\ldots$

$11.60\dot{2}5\dot{3} = 11.602\,532\,532\,53\ldots$

Decimals to fractions

A decimal can be changed to a fraction by considering place value.

…	100	10	1	•	$\frac{1}{10}$	$\frac{1}{100}$	$\frac{1}{1000}$	…
	3	4	2	•	1	6	8	

$342.168 = 300 + 40 + 2 + \frac{1}{10} + \frac{6}{100} + \frac{8}{1000}$

e.g. Change 0.58 to a fraction.

$0.58 = 0 \times 1$ and $5 \times \frac{1}{10}$ and $8 \times \frac{1}{100}$

$= 0 + \frac{5}{10} + \frac{8}{100}$

$= \frac{50}{100} + \frac{8}{100}$ Rewriting as equivalent fractions with denominators of 100.

$= \frac{58}{100}$

$= \frac{29}{50}$ Cancelling down.

Percentages

Percentages are fractions with a denominator of 100.

1% means 1 out of 100 or $\frac{1}{100}$

25% means 25 out of 100 or $\frac{25}{100}$ (= $\frac{1}{4}$ in its lowest terms)

Percentages to fractions

To change a percentage to a fraction, divide by 100.

e.g. Change 65% to a fraction.

$65\% = \frac{65}{100} = \frac{13}{20}$ Cancelling down.

e.g. Change $33\frac{1}{2}\%$ to a fraction.

$33\frac{1}{2}\% = \frac{33\frac{1}{2}}{100} = \frac{67}{200}$ Cancelling down to lowest terms with integers on the top and bottom.

Fractions to percentages

To change a fraction to a percentage, multiply by 100.

e.g. Change $\frac{1}{4}$ to a percentage.

$\frac{1}{4} = \frac{1}{4} \times 100\%$

$= 25\%$

Percentages to decimals

To change a percentage to a decimal, divide by 100.

e.g. Change 65% to a decimal.

$65\% = 65 \div 100$

$= 0.65$

Decimals to percentages

To change a decimal to a percentage, multiply by 100.

e.g. Change 0.005 to a percentage.

$0.005 = 0.005 \times 100\%$

$= 0.5\%$

NB To compare and order percentages, fractions and decimals, convert them all to percentages.

Percentage change

To work out the percentage change, work out the change and use the formula:

$$\textbf{percentage change} = \frac{\textbf{change}}{\textbf{original amount}} \times \textbf{100\%}$$

where change might be increase, decrease, profit, loss, error, etc.

Percentage of an amount

To find the percentage of an amount, find 1% of the amount and then the required amount.

e.g. An investment of £72 increases by 12%. What is the new amount of the investment?

1% of £72 $= £\frac{72}{100}$

$= £0.72$

12% of £72 $= 12 \times £0.72$

$= £8.64$

The new amount is £72 + £8.64 = £80.64

An alternative method uses the fact that after a 12% increase, the new amount is 100% of the original amount + 12% of the original amount or 112% of the original amount.

The new value of the investment is 112% of £72

1% of £72 $= £0.72$

112% of £72 $= 112 \times £0.72$

$= £80.64$ (as before)

Similarly, a decrease of 12% = 100% of the original amount – 12% of the original amount or 88% of the original amount.

Reverse percentages

To find the original amount after a percentage change, use reverse percentages.

e.g. A television is advertised at £335.75 after a price reduction of 15%. What was the original price?

£335.75 represents 85% of the original price (100% – 15%)

So 85% of the original price = £335.75

1% of the original price $= £\frac{335.75}{85} = £3.95$

100% of the original price $= 100 \times £3.95 = £395$

The original price of the television was £395.

e.g. A telephone bill costs £101.05 including VAT at $17\frac{1}{2}\%$. What is the cost of the bill without the VAT?

£101.05 represents 117.5% of the bill (100% + 17.5%)

117.5% of the bill = £101.05

1% of the bill $= £\frac{101.05}{117.5} = £0.86$

100% of the bill $= 100 \times £0.86 = £86$

The telephone bill was £86 without the VAT.

NB You should check the answer by working the numbers back the other way.

Number 1–10

1 Place the following numbers in order of size, starting with the highest.

$\frac{42}{50}$, 0.85, 0.849, 82.5%, $\frac{4}{5}$

(2 *marks – deduct* $\frac{1}{2}$ *mark for each error*)

2 Find the highest common factor (HCF) and the least common multiple (LCM) of the numbers 8 and 12.

(1 *mark each for correct* HCF *and* LCM)

3 **(a)** Change $\frac{7}{8}$ to a decimal.

(b) Give your answer in part (a) correct to 2 d.p.

(1 *mark for each part*)

4 How many yards in $\frac{1}{2}$ kilometre?

(2 *marks if fully correct*)

5 Calculate the following, giving your answers in index form where possible.

(a) $12^{11} \times 12^{15}$ **(b)** $17^4 \div 17^4$

(1 *mark for each part*)

6 What is the difference between the square root of 64 and the cube root of 64?

(1 *mark*)

7 Give an approximate value for:

$$\frac{5.18 \times \pi}{2.96 \times \sqrt{25.4}}$$

(2 *marks if fully correct*)

8 A caravan valued at £13 500 depreciates by 12% each year. What is the value of the caravan after

(a) one year **(b)** two years?

(1 *mark for each part*)

9 A weighing machine records a weight of 5 kg when the actual weight is 4.96 kg. What is the percentage error on the actual weight?

(2 *marks if fully correct*)

10 The length of a metal rod increases by 2.2% to 49.056 cm when it is heated. What was the original length of the metal rod?

(3 *marks if fully correct*)

***Total marks*: 20**

ANSWERS & TUTORIALS

1 0.85, 0.849, $\frac{42}{50}$, 82.5%, $\frac{4}{5}$ (highest to lowest)
Converting the fractions and decimals to percentages:
$\frac{42}{50} = 84\%$, $0.85 = 85\%$, $0.849 = 84.9\%$, $\frac{4}{5} = 80\%$

2 HCF = 4 Common factors are 1, 2, 4 and HCF = 4
LCM = 24 Common multiples are 24, 48, 72, … and LCM = 24

3 **(a)** 0.875 $7 \div 8 = 0.875$
(b) 0.88 (2 d.p.) Rounding up.

4 550 yards
8 km = 5 miles
= 5×1760 yards as 1 mile = 1760 yards
= 8800 yards
1 km = $8800 \div 8$ yards = 1100 yards
Dividing by 8 to find 1 km.
$\frac{1}{2}$ km = $1100 \div 2$ yards = 550 yards
Dividing by 2 to find $\frac{1}{2}$ km.

5 **(a)** 12^{26}
$12^{11} \times 12^{15} = 12^{11+15} = 12^{26}$
Using $a^m \times a^n = a^{m+n}$
(b) 1
$17^4 \div 17^4 = 17^{4-4} = 17^0 = 1$
Using $a^m \div a^n = a^{m-n}$ and $a^0 = 1$

6 Difference = 4
$\sqrt{64} = 8$ and $\sqrt[3]{64} = 4$, difference = $8 - 4 = 4$

7 1
$$\frac{5.18 \times \pi}{2.96 \times \sqrt{25.4}} \approx \frac{5 \times 3}{3 \times \sqrt{25}} = \frac{15}{15} = 1$$
As π is approximately 3 and $\sqrt{25.4}$ is approximately 5.

8 **(a)** £11 880
New value = 88% (100% – 12%) of the original value.
After one year, value is 88% of £13 500
= £11 880
(b) £10 454.40
After two years, value is 88% of £11 880
= £10 454.40

9 0.8%
$$\text{Percentage error} = \frac{\text{error}}{\text{original amount}} \times 100\%$$
Error = $5 - 4.96 = 0.04$
$$\text{Percentage error} = \frac{0.04}{4.96} \times 100\%$$
= 0.806 451 6%
= 0.8% to a reasonable degree of accuracy

10 48 cm After an increase of 2.2%, the length is 102.2% of the original length.
102.2% of original length = 49.056 cm
1% of original length = $49.056 \div 102.2$
= 0.48 cm
100% of original length = 100×0.48
= 48 cm

SCORE

TOTAL SCORE OUT OF 20

Number 1–10

1 Complete the following table.

Fraction	Decimal	Percentage
$\frac{1}{2}$		50%
	0.25	
		40%
$\frac{3}{8}$		

(2 *marks – deduct* $\frac{1}{2}$ *mark for each error*)

2 Write 2420 as a product of its prime factors.
(2 *marks if fully correct*)

3 How many $\frac{3}{4}$ litre bottles can be filled from a container holding 15 litres?
(2 *marks*)

4 How many metres in 500 yards?
(2 *marks if fully correct*)

5 Calculate the following, giving your answers in index form where possible.
(a) $18^6 \div 18^4$ **(b)** $4^3 \times 5^2$
(1 *mark for each part*)

6 Give an approximate value for:

$$\frac{61.69}{20 \times \pi} + \sqrt{81.4}$$

(1 *mark*)

7 The number of incidents handled by a coastguard increases by 9% each year. If the number of incidents last year was 500 how many casualties will there be
(a) this year **(b)** next year?
(1 *mark for each part*)

8 A store buys scarves at £8.24 each and sells them at a price of £16.99. What is the percentage profit?
(2 *marks*)

9 22% of a certain number is 770. What is the number?
(2 *marks*)

10 A car is sold for £5568 after a depreciation of 42% of the original purchase price. Calculate the original purchase price.
(3 *marks*)

***Total marks*: 20**

ANSWERS & TUTORIALS

SCORE

1

Fraction	Decimal	Percentage
$\frac{1}{2}$	0.5	50%
$\frac{1}{4}$	0.25	25%
$\frac{2}{5}$	0.4	40%
$\frac{3}{8}$	0.375	37.5%

See notes on cards 8 and 9 about converting.

2 $2 \times 2 \times 5 \times 11 \times 11$ or $2^2 \times 5 \times 11^2$
See notes on card 1 about prime factors.

3 20 bottles $15 \div \frac{3}{4} = 15 \times \frac{4}{3} = \frac{60}{3} = 20$

4 450 m
500 yards = 1500 feet 3 feet = 1 yard
= 1500 × 30 cm 1 foot = 30 cm
= 45 000 cm
= 450 m Convert to metres.

5 **(a)** 18^2 $18^6 \div 18^4 = 18^{6-4} = 18^2$
Using $a^m \div a^n = a^{m-n}$
(b) 1600
As the base numbers are not the same you cannot use the rules of indices.
$4^3 \times 5^2 = 64 \times 25 = 1600$

6 10
$$\frac{61.69}{20 \times \pi} + \sqrt{81.4} \approx \frac{60}{20 \times 3} + \sqrt{81} = 1 + 9 = 10$$
As π is approximately 3.

SCORE

7 **(a)** 545
Increase of 9% means a multiplier of 109%.
Number of incidents this year
= 109% × 500 = 545
(b) 594
Number of incidents next year
= 109% × 545 = 594.05 = 594 (to the nearest whole number)

8 106%
$$\text{Percentage profit} = \frac{\text{profit}}{\text{original price}} \times 100\%$$
Profit = £16.99 – £8.24 = £8.75
$$\text{Percentage profit} = \frac{8.75}{8.24} \times 100\%$$
$$= 106.189\,32\% = 106\%$$
(to a reasonable degree of accuracy)

9 3500
If 22% = 770
1% = 770 ÷ 22 = 35 Dividing both sides by 22.
100% = 100 × 35 = 3500 Multiplying both sides by 100.

10 £9600
After depreciation, the price represents 58% (100% – 42%) of the original purchase price.
58% = £5568
1% = £5568 ÷ 58 = £96
100% = £96 × 100 = £9600

TOTAL SCORE OUT OF 20

Ratio and proportion

A **ratio** allows one quantity to be compared to another quantity in a similar way to fractions.

e.g. In a box there are 12 lemons and 16 oranges. The ratio of lemons to oranges is 12 to 16, written as 12 : 16.

The order is important in ratios as the ratio of oranges to lemons is 16 to 12 or 16 : 12.

Equivalent ratios

Equivalent ratios are ratios that are equal to each other. The following ratios are all equivalent to 2 : 5.

$$\begin{aligned} 2:5 &= 4:10 \\ &= 6:15 \\ &= 8:20 \\ &= \dots \end{aligned}$$

Equal ratios can be found by multiplying or dividing both sides of the ratio by the same number.

e.g. Express the ratio 40p to £2 in its simplest form.

You must ensure that the units are the same.
Remember £2 = 200p.
The ratio is 40 : 200 = 1 : 5 in its simplest form.
Dividing both sides of the ratio by 40.

e.g. Two lengths are in the ratio 4 : 5. If the first length is 60 cm, what is the second length?

The ratio is $4:5 = 4\text{ cm} : 5\text{ cm}$
$= 1\text{ cm} : \frac{5}{4}\text{ cm}$
Writing as an equivalent ratio with 1 cm on the left-hand side.
$= 60\text{ cm} : 60 \times \frac{5}{4}\text{ cm}$
Writing as an equivalent ratio with 60 cm on the left-hand side.
$= 60\text{ cm} : 75\text{ cm}$

So the second length is 75 cm.

Proportional parts

To share an amount into **proportional** parts, add up the individual parts and divide the amount by this number to find the value of one part.

e.g. £50 is to be divided between two sisters in the ratio 3 : 2. How much does each get?

Number of parts = 3 + 2
= 5
Value of each part = £50 ÷ 5
= £10
The two sisters receive £30 (3 parts at £10 each) and £20 (2 parts at £10 each).

NB Check that the amounts add up correctly (i.e. £30 + £20 = £50).

Standard form

Standard form is a short way of writing very large and very small numbers. Standard form numbers are always written as:

$$\mathbf{A \times 10^n}$$

where A lies between 1 and 10 and n is a natural number.

Very large numbers

e.g. Write 267 000 000 in standard form.
Write down 267 000 000
then place the decimal point so A lies between 1 and 10.

2.6 7 0 0 0 0 0 0 0

To find n, count the 'power of 10'.
Here, $n = 8$ so $267\,000\,000 = 2.67 \times 10^8$

Very small numbers

e.g. Write 0.000 000 231 in standard form.
Write down .000 000 321
then place the decimal point so A lies between 1 and 10.

0 0 0 0 0 0 0 3.21

To find n in 0.000 000 321, count the 'power of 10'.
Here, $n = {}^-7$ so $0.000\,000\,321 = 3.21 \times 10^{-7}$

Adding and subtracting

To add (or subtract) numbers in standard form *when the powers are the same* you can proceed as follows.

e.g. $(4.8 \times 10^{11}) + (3.1 \times 10^{11})$
$= (4.8 + 3.1) \times 10^{11} = 7.9 \times 10^{11}$

e.g. $(4.63 \times 10^{-2}) - (2.7 \times 10^{-2})$
$= (4.63 - 2.7) \times 10^{-2} = 1.93 \times 10^{-2}$

To add (or subtract) numbers in standard form *when the powers are **not** the same*, convert the numbers to ordinary form.

e.g. $(8.42 \times 10^6) + (6 \times 10^7)$
$= 8\,420\,000 + 60\,000\,000$ Converting.
$= 68\,420\,000 = 6.842 \times 10^7$

Multiplying and dividing

To multiply (or divide) numbers in standard form, use the rules of indices.

e.g. $(7.5 \times 10^4) \times (3.9 \times 10^7)$
$= (7.5 \times 3.9) \times (10^4 \times 10^7)$ Collecting powers of 10.
$= 29.25 \times 10^{4+7}$ Using rules of indices.
$= 29.25 \times 10^{11}$
$= 2.925 \times 10^{12}$ As $29.25 = 2.925 \times 10^1$.

e.g. $(3 \times 10^5) \div (3.75 \times 10^8)$
$= (3 \div 3.75) \times (10^5 \div 10^8)$ Collecting powers of 10.
$= 0.8 \times 10^{5-8}$ Using rules of indices.
$= 0.8 \times 10^{-3}$
$= 8 \times 10^{-4}$ As $0.8 = 8 \times 10^{-1}$.

Compound measures

Compound measures involve more than one unit, such as **speed** (distance and time) or **density** (mass and volume).

Speed

The formula for speed is

$$\textbf{speed} = \frac{\textbf{distance}}{\textbf{time}}$$

e.g. A taxi travels 16 miles in 20 minutes. What is the speed in miles per hour?

As the speed is measured in miles per hour, express the distance in miles and the time in hours.

Time = 20 minutes = $\frac{1}{3}$ hour

$$\text{Speed} = \frac{\text{distance}}{\text{time}} = \frac{16}{\frac{1}{3}}$$

$$= 48 \text{ mph}$$

The formula for speed can be rearranged:

distance = speed × **time** or

$$\textbf{time} = \frac{\textbf{distance}}{\textbf{speed}}$$

e.g. A cyclist travels 3.6 km at an average speed of 8 kilometres per hour. How long does the journey take?

$$\text{Time} = \frac{\text{distance}}{\text{speed}} = \frac{3.6}{8}$$

$$= 0.45 \text{ hours}$$

Remember that 0.45 hours is not 45 minutes as there are 60 minutes in one hour.

To convert hours to minutes, multiply by 60.

0.45 hours = 0.45 × 60 minutes = 27 minutes

The journey takes 27 minutes.

Density

The formula for density is:

$$\textbf{density} = \frac{\textbf{mass}}{\textbf{volume}}$$

e.g. A piece of lead weighing 170 g has a volume of 15 cm^3. Give an estimate for the density of lead.

$$\text{Density} = \frac{\text{mass}}{\text{volume}} = \frac{170}{15}$$

$$= 11.3 \text{ g/cm}^3 \text{ (3 s.f.)}$$

The formula for density can be rearranged:

mass = density × **volume** or

$$\textbf{volume} = \frac{\textbf{mass}}{\textbf{density}}$$

Simple and compound interest

With **simple interest**, the amount of interest paid is not reinvested.

With **compound interest**, the amount of interest paid is reinvested and earns interest itself.

Simple interest formula: $\boldsymbol{I = \frac{PRT}{100}}$ and $\boldsymbol{A = P + \frac{PRT}{100}}$

Compound interest formula: $\boldsymbol{A = P\left(1 + \frac{R}{100}\right)^T}$

where I = simple interest, A = total amount,
P = principal or original investment,
R = rate (% per annum, or p.a.),
T = time (in years)

e.g. £4000 is invested for 3 years at $4\frac{1}{2}$% p.a. Calculate the simple interest and the total amount.

Using the formula $I = \frac{PRT}{100}$

where P = principal = £4000
R = rate = $4\frac{1}{2}$% or 4.5%
T = time = 3 years

$$I = \frac{4000 \times 4.5 \times 3}{100} = £540$$

$A = P + I$ Amount = principal + interest.

A = £4000 + £540 = £4540

The simple interest is £540 and the total amount is £4540.

e.g. £2500 is invested at 6.5% p.a. compound interest. What is the amount after 2 years?

Compound interest can be found by repeatedly applying the simple interest formula.

$$A = P + \frac{PRT}{100}$$

where P = principal = £2500
R = rate = 6.5%
T = time = 1 year for each year

Year 1 $\quad A = 2500 + \frac{2500 \times 6.5 \times 1}{100}$
$= 2500 + 162.5$
$= £2662.50$ Writing £2662.5 as £2662.50

Year 2 $\quad A = 2662.5 + \frac{2662.5 \times 6.5 \times 1}{100}$ As P = £2662.50 now.
$= 2662.5 + 173.0625$
$= 2835.5625$
$= £2835.56$ correct to the nearest penny.

Alternatively, using the compound interest formula:

$$A = P\left(1 + \frac{R}{100}\right)^T$$

$A = 2500\left(1 + \frac{6.5}{100}\right)^2$ As T = 2 years.
$= 2835.5625$
$= £2835.56$ correct to the nearest penny.

Number 11–14

1 The distance from the earth to the sun is 9.3×10^7 miles. Write this number in ordinary form.

(2 *marks*)

2 The weight of a hydrogen atom is written as 0.000 000 000 000 000 000 001 67 milligrams. Write this number in standard form.

(2 *marks*)

3 A plan of an estate is drawn on a scale of 1 : 50. On the plan, a fence is 35 cm long. How long is the actual fence in metres?

(2 *marks*)

4 Three children aged 2 years, 3 years and 7 years old are given a sum of £60 to be divided in the ratio of their ages.

(a) How much does each child receive?

(b) How much would each child receive if the same amount of money had been divided a year later?

(2 *marks for part* (*a*) *and* 1 *mark for part* (*b*))

5 A van travels 14 miles in 20 minutes. What is the speed in miles per hour?

(2 *marks*)

6 Calculate the difference in interest earned on £4000 over 2 years at:

(a) 6.5% p.a. simple interest

(b) 6.5% p.a. compound interest.

(2 *marks if completely correct*)

7 The value of m is 5×10^{-2} and of n is 2.5×10^{-3}. Without using a calculator find:

(a) the value of mn giving your answer in standard form

(b) the value of $m + n$.

(2 *marks for each part*)

8 A coach travels 45 miles at 60 mph on a motorway and 20 miles at 25 mph on a country road. What is the average speed over the whole of the journey?

(3 *marks if fully correct*)

***Total marks* = 20**

ANSWERS & TUTORIALS

1 93 000 000 miles
See notes on card 12 about standard form.

2 1.67×10^{-21} milligrams
See notes on card 12 about standard form.

3 17.5 m
$1 : 50 = 35 : 50 \times 35 = 35 : 1750$
so the pathway is 1750 cm = 17.5 metres

4 **(a)** £10, £15 and £35
Their ages are in the ratio 2 : 3 : 7.
Number of parts = 2 + 3 + 7 = 12
Value of each part = £60 ÷ 12 = £5
So they receive £10 (2 parts), £15 (3 parts) and £35 (7 parts).

(b) £12, £16 and £32
The following year their ages are in the ratio 3 : 4 : 8.
Number of parts = 3 + 4 + 8 = 15
Value of each part = £60 ÷ 15 = £4
So they receive £12 (3 parts), £16 (4 parts) and £32 (8 parts).

5 42 miles per hour
To express speed in miles per hour, the time must be in hours so 20 minutes = $\frac{1}{3}$ hour.

$$\text{Speed} = \frac{\text{distance}}{\text{speed}} = \frac{14}{\frac{1}{3}} = 42 \text{ mph}$$

6 **(a)** £520

$$I = \frac{P \times R \times T}{100} = \frac{4000 \times 6.5 \times 2}{100} = £520$$

(b) £536.90, difference = £16.90

$$A = P\left(1 + \frac{R}{100}\right)^T = 4000\left(1 + \frac{6.5}{100}\right)^2$$
$$= £4536.90$$
$$I = £4536.90 - £4000 = £536.90$$

Difference = £536.90 – £520 = £16.90

7 **(a)** 1.25×10^{-4}

$$mn = (5 \times 10^{-2}) \times (2.5 \times 10^{-3})$$
$$= 5 \times 2.5 \times 10^{-2} \times 10^{-3}$$
$$= 12.5 \times 10^{-2 + -3}$$
$$= 12.5 \times 10^{-5}$$
$$= 1.25 \times 10^{-4} \quad \text{As } 12.5 = 1.25 \times 10^{1}$$

(b) 0.0525 or 5.25×10^{-2}

$$m + n = 5 \times 10^{-2} + 2.5 \times 10^{-3}$$
$$= 0.05 + 0.0025$$

Writing the numbers in full.

$$= 0.0525 \text{ or } 5.25 \times 10^{-2}$$

8 41.9 mph (3 s.f.)
Total distance = 45 + 20 = 65 miles

$$\text{Time for 45 miles at 60 mph} = \frac{45}{60} = 0.75 \text{ hours}$$

$$\text{Time for 20 miles at 25 mph} = \frac{20}{25} = 0.8 \text{ hours}$$

Total time = 0.75 + 0.8 = 1.55 hours

$$\text{Total speed} = \frac{\text{distance}}{\text{time}} = \frac{65}{1.55}$$
$$= 41.935\,484 = 41.9 \text{ mph (3 s.f.)}$$

SCORE

SCORE

TOTAL SCORE OUT OF 20

Number 11–14

1 The circulation figure for a newspaper is $10\frac{1}{2}$ million. Write this number in standard form.
(2 *marks*)

2 Express 8.45×10^{-4} as an ordinary number.
(2 *marks*)

3 Express the ratio 3 km to 600 m in its simplest form.
(2 *marks*)

4 Two villages with populations of 550 and 160 receive a grant for £3550. The councils agree to share the money in proportion to the population. How much does each village get?
(1 *mark for each correct answer*)

5 A piece of lead weighing 226 g has a volume of 20 cm^3. Give an estimate for the density of lead.
(2 *marks*)

6 A sum of £2000 is invested at 5.5% p.a. simple interest. How long will it be before the amount equals £2275?
(3 *marks*)

7 What is the compound interest on £10 000 invested over 2 years at 5.75% p.a?
(3 *marks*)

8 Light travels at 2.998×10^8 m/s. Calculate how far light travels in one year, giving your answer in metres and using standard index form.
(4 *marks if fully correct*)

***Total marks* = 20**

ANSWERS & TUTORIALS

1 1.05×10^7
Using the fact that 1 million = 10^6
10.5 million = $10.5 \times 10^6 = 1.05 \times 10^7$

2 0.000 845
See notes on card 12 about standard form.

3 5 : 1
The ratio 3 km to 600 m = 3000 : 600 = 5 : 1
Converting 3 km to 3000 m so that both sides of the ratio are in the same units.

4 £2750 and £800
The grant is shared in the ratio of 550 to 160.
Number of parts = 550 + 160 = 710
Value of each part = £3550 ÷ 710 = £5
The villages get £2750 (550 × £5) and £800 (160 × £5).

5 11.3 g/cm³

$$\text{Density} = \frac{\text{mass}}{\text{volume}} = \frac{226}{20} = 11.3\ \text{g/cm}^3$$

6 2.5 years or 2 years 6 months
The amount $A = P + I$ and
$I = £2275 - £2000 = £275$

$$I = 275 = \frac{PRT}{100}$$

$$275 = \frac{2000 \times 5.5 \times T}{100}$$

Substituting $P = £2000$ and $R = 5.5\%$.

$$275 = 110 \times T$$

$$T = \frac{275}{110} = 2.5$$

The time is 2.5 years or 2 years and 6 months.

7 £1183.06 (nearest penny)

$$A = P\left(1 + \frac{R}{100}\right)^T = 10\,000\left(1 + \frac{5.75}{100}\right)^2$$

Substituting $P = £10\,000$, $R = 5.75\%$ and $T = 2$.
$A = £11\,183.0625$
Compound interest
= £11 183.062 5 – £10 000.00
= £1183.06 (to the nearest penny)

8 9.45×10^{15} metres (to an appropriate degree of accuracy)
If light travels 2.998×10^8 metres in one second then it travels:
$2.998 \times 10^8 \times 60$ metres in one minute
$= 2.998 \times 10^8 \times 60 \times 60$ metres in one hour
$= 2.998 \times 10^8 \times 60 \times 60 \times 24$ metres in one day
$= 2.998 \times 10^8 \times 60 \times 60 \times 24 \times 365$ metres in a year
$= 9.454\,492\,8 \times 10^{15}$ metres in a year
$= 9.45 \times 10^{15}$ metres (to an appropriate degree of accuracy)

SCORE

SCORE

TOTAL SCORE OUT OF 20

Rational and irrational numbers

A **rational number** is one which can be expressed in the form $\frac{p}{q}$ where p and q are integers.

Rational numbers include $\frac{1}{5}$, $0.\dot{3}$, 7, $\sqrt{9}$, $\sqrt[3]{64}$, etc.
Irrational numbers include $\sqrt{2}$, $\sqrt{3}$, $\sqrt[3]{20}$, π, π^2, etc.

Irrational numbers involving square roots are also called **surds**. Surds can be multiplied and divided according to the following rules.

$$\sqrt{a} \times \sqrt{b} = \sqrt{a \times b} \qquad \frac{\sqrt{a}}{\sqrt{b}} = \sqrt{\frac{a}{b}}$$

e.g. $\sqrt{3} \times \sqrt{3} = \sqrt{3 \times 3} = \sqrt{9} = 3$

$\sqrt{2} \times \sqrt{8} = \sqrt{2 \times 8} = \sqrt{16} = 4$

$\frac{\sqrt{48}}{\sqrt{12}} = \sqrt{\frac{48}{12}} = \sqrt{4} = 2$

Similarly $\sqrt{50} = \sqrt{25 \times 2} = \sqrt{25} \times \sqrt{2} = 5 \times \sqrt{2} = 5\sqrt{2}$

As it is usual to write $5 \times \sqrt{2}$ as $5\sqrt{2}$.

Also $\sqrt{5} + \sqrt{45} = \sqrt{5} + \sqrt{9 \times 5} = \sqrt{5} + 3\sqrt{5} = 4\sqrt{5}$

Recurring decimals

Recurring decimals are rational numbers as they can all be expressed as fractions in the form $\frac{p}{q}$ where p and q are integers.

e.g. 0.166 666 666… written $0.1\dot{6} = \frac{1}{6}$

0.142 857 142 857… written $0.\dot{1}42\,85\dot{7} = \frac{1}{7}$

0.272 727 27… written $0.\dot{2}\dot{7} = \frac{3}{11}$

e.g. Change $0.\dot{8}$ to a fraction.

$10 \times 0.\dot{8} = 8.888\,888\,8\ldots$ Multiplying both sides by 10.

and $1 \times 0.\dot{8} = 0.888\,888\,8\ldots$ Then subtracting …

$9 \times 0.\dot{8} = 8$ $8.888\,888\,8\ldots - 0.888\,888\,8\ldots$

and $0.\dot{8} = \frac{8}{9}$ Dividing both sides by 9.

e.g. Convert $14.\dot{2}\dot{3}$ to a mixed number.

$100 \times 14.\dot{2}\dot{3} = 1423.232\,323\ldots$ Multiplying both sides by 100.

and $1 \times 14.\dot{2}\dot{3} = 14.232\,323\ldots$ Then subtracting …

$99 \times 14.\dot{2}\dot{3} = 1409$ $1423.232\,323\ldots - 14.232\,323\ldots$

and $14.\dot{2}\dot{3} = \frac{1409}{99}$ Dividing both sides by 99.

$= 14\frac{23}{99}$ As a mixed number.

NB You could also write $14.\dot{2}\dot{3}$ as $14 + 0.\dot{2}\dot{3}$ and convert $0.\dot{2}\dot{3}$ to a fraction before rewriting as a mixed number.

Fractional indices

You should know that, in general:

$$\boldsymbol{a^m \times a^n = a^{m+n}}$$

$$\boldsymbol{a^m \div a^n = a^{m-n}}$$

$$\boldsymbol{a^{-m} = \frac{1}{a^m}}$$

$$\boldsymbol{a^1 = a}$$

$$\boldsymbol{a^0 = 1}$$

Using these rules, you can see that:

$a^{\frac{1}{2}} \times a^{\frac{1}{2}} = a^{\frac{1}{2}+\frac{1}{2}} = a^1$

$= a$

and $a^{\frac{1}{3}} \times a^{\frac{1}{3}} \times a^{\frac{1}{3}} = a^{\frac{1}{3}+\frac{1}{3}+\frac{1}{3}}$

$= a^1$

$= a$

So any number raised to the power $\frac{1}{2}$ means $\sqrt[2]{\ }$ or $\sqrt{\ }$

i.e. $\boldsymbol{a^{\frac{1}{2}} = \sqrt{a}}$

and any number raised to the power $\frac{1}{3}$ means $\sqrt[3]{\ }$

i.e. $\boldsymbol{a^{\frac{1}{3}} = \sqrt[3]{a}}$

Similarly:

$$\boldsymbol{a^{\frac{1}{n}} = \sqrt[n]{a}}$$

e.g. $25^{\frac{1}{2}} = 5$

NB $\sqrt{25} = {}^-5$ as well but it is usual to take the positive square root.

Also, $64^{\frac{1}{3}} = \sqrt[3]{64} = 4$ and

$32^{\frac{1}{5}} = \sqrt[5]{32} = 2$

e.g. $81^{\frac{3}{4}} = (\sqrt[4]{81})^3 = 3^3 = 27$ As $81^{\frac{1}{4}} = 3$

Alternatively, you can use

$81^{\frac{3}{4}} = (\sqrt[4]{81})^3$

$= \sqrt[4]{531\,441}$

$= 27$

although this method is rather longwinded.

e.g. Work out $125^{\frac{2}{3}}$.

$125^{\frac{2}{3}} = (125^{\frac{1}{3}})^2$

$= 5^2$

$= 25$

e.g. Work out $125^{-\frac{2}{3}}$.

$125^{-\frac{2}{3}} = \frac{1}{125^{\frac{2}{3}}}$

$= \frac{1}{5^2}$ As $125^{\frac{2}{3}} = 5^2$

$= \frac{1}{25}$ As $5^2 = 25$

Direct and inverse proportion

With **direct proportion**, as one variable increases the other increases, and as one variable decreases the other decreases.

With **inverse proportion**, as one variable increases the other decreases, and as one variable decreases the other increases.

If y is proportional to x then you can write $y \propto x$ or else $y = kx$.

If y is inversely proportional to x then you can write $y \propto \frac{1}{x}$ or else $y = \frac{k}{x}$.

The value of k is a constant and is called the **constant of proportionality**.

NB The term 'varies as' is also used instead of 'is proportional to'.

e.g. If a is proportional to the cube of b and $a = 4$ when $b = 2$, find the value of k (the constant of proportionality) and the value of a when $b = 5$.

If a is proportional to the cube of b then

$a \propto b^3$ or $a = kb^3$.

Since $a = 4$ when $b = 2$,

then $4 = k \times 2^3$

$4 = k \times 8$

$k = \frac{1}{2}$

The equation is $a = \frac{1}{2}b^3$.

When $b = 5$ then $a = \frac{1}{2} \times 5^3$

$= \frac{1}{2} \times 125 = 62.5$

e.g. Given that p varies inversely as the cube of q and $p = 1$ when $q = \frac{1}{2}$, find the value of k (the constant of proportionality) and the value of p when $q = 3$.

As p varies inversely as q^3, then

$$p \propto \frac{1}{q^3} \text{ or } p = \frac{k}{q^3}$$

Since $p = 1$ when $q = \frac{1}{2}$, then

$$1 = \frac{k}{\frac{1}{2}^3} \text{ and } k = \frac{1}{8}$$

The equation is

$$p = \frac{1}{8q^3}$$

When $q = 3$ then

$$p = \frac{1}{8 \times 3^3} = \frac{1}{8 \times 27} = \frac{1}{216}$$

Upper and lower bounds

If a length is given as 10 cm to the nearest cm then the actual length will lie in the interval 9.5 cm to 10.499 999... cm as all values in this interval will be rounded off to 10 cm to the nearest cm. The length 10.499 999... cm is usually written as 10.5 cm although it is accepted that 10.5 cm would be rounded to 11 cm (to the nearest cm).

The value 9.5 cm is called the **lower bound** as it is the lowest value which would be rounded to 10 cm while 10.5 cm is called the **upper bound**.

e.g. A rectangle measures 10 cm by 6 cm where each measurement is given to the nearest cm. Write down an interval approximation for the area of the rectangle.

Lower bound (minimum area) $= 9.5 \times 5.5$
$= 52.25\ \text{cm}^2$

Upper bound (maximum area) $= 10.5 \times 6.5$
$= 68.25\ \text{cm}^2$

The interval approximation is $52.25 - 68.25\ \text{cm}^2$

e.g. Calculate the lower and upper bounds of:

	lower bound	upper bound
8000 given to the nearest 1000	7500	8500
250 given to the nearest 10	245	255
4.13 given to 2 decimal places	4.125	4.135
20.0 given to 3 significant figures	19.95	20.05

e.g. The value of p is 215 and the value of q is 5 with both figures being given to the nearest whole number. Calculate the maximum and minimum values of:

$$p + q,\ p - q,\ p \times q,\ p \div q$$

Using $p_{min} = 214.5$ $\quad p_{max} = 215.5$
$q_{min} = 4.5$ $\quad q_{max} = 5.5$

$p + q$ maximum $= 215.5 + 5.5 = 221$
minimum $= 214.5 + 4.5 = 219$

$p - q$ maximum $= 215.5 - 4.5 = 211$
minimum $= 214.5 - 5.5 = 209$

NB To get the maximum value of $p - q$ you need to work out $p_{max} - q_{min}$ and to get the minimum value of $p - q$ you need to work out $p_{min} - q_{max}$.

$p \times q$ maximum $= 215.5 \times 5.5 = 1185.25$
minimum $= 214.5 \times 4.5 = 965.25$

$p \div q$ maximum $= 215.5 \div 4.5 = 47.888\,888$
minimum $= 214.5 \div 5.5 = 39$

NB To get the maximum value of $p \div q$ you need to work out $p_{max} \div q_{min}$ and to get the minimum value of $p \div q$ you need to work out $p_{min} \div q_{max}$.

Check yourself 5

Number 15–18 (higher tier)

1 Which of the following numbers are rational?

$(\sqrt{2})^3$ $16^{\frac{1}{2}}$ $\pi + 1$ 3^{-1} $\sqrt{6\frac{1}{4}}$

(*2 marks – deduct* $\frac{1}{2}$ *mark for each error*)

2 Simplify the following expressions, leaving your answers in surd form.

(a) $\sqrt{5} \times \sqrt{20}$ **(b)** $\sqrt{5} + \sqrt{20}$

(*1 mark for each part*)

3 Write $\frac{\sqrt{3}}{\sqrt{2}}$ in the form

$\frac{\sqrt{a}}{\sqrt{b}}$

where a and b are integers.

(*1 mark*)

4 Calculate a fraction equivalent to $0.\dot{3}4\dot{7}$.

(*2 marks*)

5 Write down the following in order of size with the smallest first.

$243^{\frac{3}{5}}$ 10^{-1} 250^0 $1024^{\frac{3}{5}}$ $(-5)^3$ 5^{-3}

(*2 marks – deduct* $\frac{1}{2}$ *mark for each error*)

6 If $x = \sqrt{8} + \sqrt{2}$ show that $x^2 = 18$ without using your calculator.

(*3 marks if fully correct*)

7 The braking distance, b, of a car is proportional to the square of the speed, s. If the braking distance for a car travelling at 50 mph is 125 feet, find:

(a) the braking distance when the speed is 70 mph

(b) the speed when the braking distance is 45 feet.

(*3 marks if fully correct*)

8 A runner completes a sprint in 12.31 seconds to the nearest one hundredth of a second. What is the shortest time that it could actually be?

(*2 marks*)

9 A train travels 45 miles (to the nearest mile) in a time of 40 minutes (to the nearest minute). What is the maximum and minimum speed of the train in miles per hour to 3 significant figures?

(*3 marks if fully correct*)

***Total marks* = 20**

SCORE

1 $16^{\frac{1}{2}}$ 3^{-1} $\sqrt{6\frac{1}{4}}$

$16^{\frac{1}{2}} = 4$ as $16^{\frac{1}{2}} = \sqrt{16} = 4$

$3^{-1} = \frac{1}{3}$ as $3^{-1} = \frac{1}{3^1} = \frac{1}{3}$

$\sqrt{6\frac{1}{4}} = \frac{5}{2}$ as $= \sqrt{6\frac{1}{4}} = \sqrt{\frac{25}{4}} = \frac{\sqrt{25}}{\sqrt{4}} = \frac{5}{2}$

2 **(a)** 10 $\sqrt{5} \times \sqrt{20} = \sqrt{100} = 10$

(b) $3\sqrt{5}$ $\sqrt{5} + \sqrt{20} = \sqrt{5} + \sqrt{4} \times \sqrt{5}$

$= \sqrt{5} + 2\sqrt{5} = 3\sqrt{5}$

3 $\frac{\sqrt{3}}{\sqrt{2}} = \frac{\sqrt{3}}{\sqrt{2}} \times \frac{\sqrt{2}}{\sqrt{2}} = \frac{\sqrt{3 \times 2}}{\sqrt{2 \times 2}} = \frac{\sqrt{6}}{2}$

4 $\frac{347}{999}$

$1000 \times 0.\dot{3}4\dot{7} = 347.347347\ldots$ Multiplying both sides by 1000.

$1 \times 0.\dot{3}4\dot{7} = 0.347347\ldots$

$999 \times 0.\dot{3}4\dot{7} = 347$ Subtracting,

So $0.\dot{3}4\dot{7} = \frac{347}{999}$ Dividing by 999.

5 Order is $(^-5)^3$, 5^{-2}, 10^{-1}, 250^0, $243^{\frac{3}{5}}$, $1024^{\frac{3}{5}}$

$243^{\frac{3}{5}} = (243^{\frac{1}{5}})^3 = 3^3 = 27$

$10^{-1} = \frac{1}{10}$ As $a^{-n} - \frac{1}{a^n}$

$250^0 = 1$ As $a^0 = 1$

$1024^{\frac{3}{5}} = (1024^{\frac{1}{5}})^3 = 4^3 = 64$

$(^-5)^3 = {}^-5 \times {}^-5 \times {}^-5 = {}^-125$

$5^{-2} = \frac{1}{5^2} = \frac{1}{25}$

SCORE

6 If $x = \sqrt{8} + \sqrt{2}$ then

$x^2 = (\sqrt{8} + \sqrt{2})(\sqrt{8} + \sqrt{2})$

$= 8 + 2\sqrt{8}\sqrt{2} + 2 = 8 + 2\sqrt{16} + 2$

$= 8 + 2 \times 4 + 2 = 18$ as required

7 **(a)** 245 feet

$b = ks^2$

$125 = k \times 50^2$

$k = 0.05$ so $b = 0.05s^2$

When $s = 70$, $b = 0.05 \times 70^2 = 245$ feet

(b) 30 mph

When $b = 45$, $45 = 0.05 \times s^2$

$s^2 = 900$ $s = 30$ mph

8 12.305 seconds

If 12.31 seconds is expressed correct to the nearest one hundredth of a second then the time is $12.31 \pm (0.01 \div 2) = 12.31 \pm (0.005)$.

The shortest time $= 12.31 - 0.005 = 12.305$

9 Maximum = 69.1 mph (3 s.f.),

Minimum = 65.9 mph (3 s.f.)

$\text{Distance}_{min} = 44.5$ miles

$\text{Distance}_{max} = 45.5$ miles

$\text{Time}_{min} = 39.5$ minutes

$\text{Time}_{max} = 40.5$ minutes

$\text{Speed}_{max} = \frac{45.5}{\frac{39.5}{60}} = 69.1$ mph (3 s.f.)

$\text{Speed}_{min} = \frac{44.5}{\frac{40.5}{60}} = 65.9$ mph (3 s.f.)

TOTAL

Expressions

An **algebraic expression** is a collection of algebraic quantities along with their + and – signs.

Substitution

Substitution means replacing the letters in an expression (or formula) by given numbers.

e.g. When $a = 3$, $b = 2$ and $c = 5$ then:

$a + b + c = 3 + 2 + 5 = 10$

$a \times b \times c = 3 \times 2 \times 5 = 30$

$\frac{b^2 + 2a}{c} = \frac{2^2 + 2 \times 3}{5} = \frac{10}{5} = 2$

etc.

Simplifying

Like terms are numerical multiples of the same algebraic quantity.

For example, $3a$, ^-5a, a and $\frac{1}{2}a$ are all like terms because they are multiples of the same algebraic quantity i.e. a.

Similarly, x^2y, $5x^2y$, $^-20x^2y$ and $\frac{x^2y}{5}$ are all like terms because they are multiples of the same algebraic quantity i.e. x^2y.

Adding and subtracting terms

You can add or subtract like terms. The process of adding and subtracting like terms in an expression (or an equation) is called **simplifying**.

e.g. Simplify the following expressions.

$3x + 3y + 5x - 7y$
$= 3x + 5x + 3y - 7y$ — Collecting together like terms.
$= 8x - 4y$

$5p - (4q - 2p)$
$= 5p - 4q + 2p$ — As $- \times - = +$ for the brackets.
$= 7p - 4q$

$4ab + 6a - 2b + 5ba$
$= 4ab + 5ab + 6a - 2b$ — As $5ba$ is the same as $5ab$.
$= 9ab + 6a - 2b$

Multiplying and dividing terms

The process of multiplying or dividing terms is also called simplifying. When dividing terms, they can be simplified by **cancelling**.

e.g. Simplify the following expressions.

$3p \times 4q = 3 \times 4 \times p \times q = 12pq$

$5a \times 7a = 5 \times 7 \times a \times a = 35a^2$ — As $a \times a = a^2$.

$8ab^2 \div 2ab = \frac{8ab^2}{2ab} = \frac{^4\cancel{8}\cancel{a}b^{\cancel{2}}}{\cancel{2}\cancel{a}\cancel{b}} = 4b$ — Cancelling top and bottom.

Algebraic indices

From Number card 3, in general:

$$\boldsymbol{a^m \times a^n = a^{m+n}}$$

$$\boldsymbol{a^m \div a^n = a^{m-n}}$$

$$\boldsymbol{a^{-m} = \frac{1}{a^m}}$$

and $\boldsymbol{a^0 = 1}$

The same laws of indices apply to algebra.

$p^2 \times p^5 = p^{2+5}$

$\quad = p^7$

$q^7 \div q^7 = q^{7-7}$

$\quad = q^0$

$\quad = 1$ As $q^0 = 1$.

$r^4 \times r \times r^{11} = r^{4+1+11}$ As $r = r^1$.

$\quad = r^{16}$

$s^3 \div s^5 = s^{3-5} = s^{-2} = \frac{1}{s^2}$ As $s^{-2} = \frac{1}{s^2}$.

Expanding and factorising

Brackets are used to group algebraic terms and the process of removing the brackets is called **expanding**. The process of rewriting an expression including the brackets is called **factorising**.

Expanding brackets

When expanding brackets you must multiply all of the terms inside the brackets by the term outside.

e.g. Expand the following.

$3(p - 7q) = 3 \times p + 3 \times {}^{-}7q$

$\quad = 3p - 21q$

${}^{-}6(a - 2b) = {}^{-}6 \times a - 6 \times {}^{-}2b$

$\quad = {}^{-}6a + 12b$

e.g. Expand and simplify the following.

$2a - 3(4b - 3a) = 2a - 3 \times 4b - 3 \times {}^{-}3a$

$\quad = 2a - 12b + 9a$

$\quad = 11a - 12b$

Factorising into brackets

To factorise an expression, look for terms that have common factors.

e.g. Factorise the following.

$10a - 15 = 5(2a - 3)$

$9rs + 12st = 3s(3r + 4t)$

$8ab^2 - 16a^2b = 8ab(b - 2a)$

NB You should check your answers by expanding the brackets.

Binomial expressions

A **binomial expression** consists of two terms such as $(a + b)$ or $(5x - 2z)$.

To **expand** the product of two binomial expressions, multiply *each term* in the first expression by *each term* in the second expression.

$$\begin{aligned}(a + b)(c + d) &= a(c + d) + b(c + d)\\ &= a \times c + a \times d + b \times c + b \times d\\ &= ac + ad + bc + bd\end{aligned}$$

This can be shown in this area diagram.

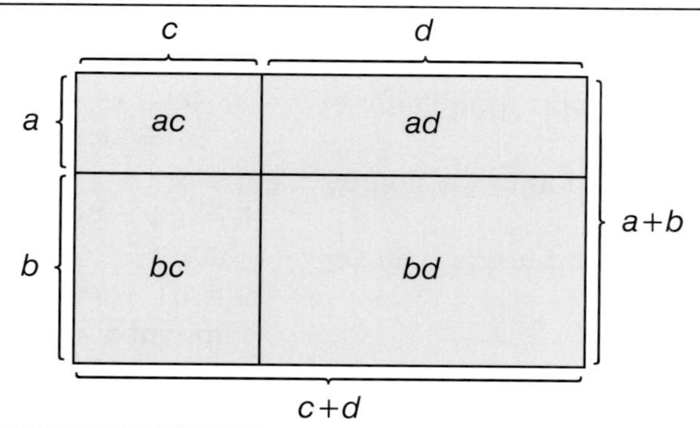

The area of $(a + b)(c + d)$ is the same as the total area of $ac + ad + bc + bd$.
So $(a + b)(c + d) = ac + ad + bc + bd$

The word **FOIL** can also be used as a reminder of how to expand these binomial expressions:

F = First	$(a + b)(c + d)$	$a \times c$
O = Outsides	$(a + b)(c + d)$	$a \times d$
I = Insides	$(a + b)(c + d)$	$b \times c$
L = Last	$(a + b)(c + d)$	$b \times d$

Expanding and factorising

Use the reverse process to write a **quadratic** as a product of brackets.

e.g. Factorise $x^2 - 6x + 5$.

Write $x^2 - 6x + 5 = (x \quad)(x \quad)$ As $x \times x = x^2$.

Then search for two numbers that multiply together to give the $^{+}5$.

Try substituting:

$^{+}\mathbf{1} \times {}^{+}\mathbf{5}$ gives $^{+}5$ $(x + \mathbf{1})(x + \mathbf{5}) = x^2 + 6x + 5$

$^{-}\mathbf{1} \times {}^{-}\mathbf{5}$ gives $^{+}5$ $(x - \mathbf{1})(x - \mathbf{5}) = x^2 - 6x + 5$ ✓

NB These are the only pairs of numbers that multiply together to give $^{+}5$. With practice, you should find the numbers quite quickly.

Solving equations

The equals sign in an **algebraic equation** provides a balance between the two sides. To maintain this balance, always make sure that whatever you do to one side of the equation you also do to the other side of the equation.

e.g. Solve $x + 10 = 5$.
$x = {}^-5$ Taking 10 from both sides.

e.g. Solve $x - 4.5 = 2$.
$x = 6.5$ Adding 4.5 to both sides.

e.g. Solve $4x = 14$.
$x = 3\frac{1}{2}$ Dividing both sides by 4.

e.g. Solve $\frac{x}{3} = 7$.
$x = 21$ Multiplying both sides by 3.

e.g. Solve $6(3x - 5) = 42$.
$6 \times 3x + 6 \times {}^-5 = 42$ Expanding brackets.
$18x - 30 = 42$
$18x = 72$ Adding 30 to both sides.
$x = 4$ Dividing both sides by 18.

Alternatively:
$6(3x - 5) = 42$
$3x - 5 = 7$ Dividing both sides by 6.
$3x = 12$ Adding 5 to both sides.
$x = 4$ Dividing both sides by 3.

Solving inequalities

Inequalities can be solved in exactly the same way as equalities (i.e. **equations**) except that when multiplying or dividing by a negative number you must reverse the inequality sign.

e.g. $4y + 6 < 26$
$4y < 20$ Taking 6 from both sides.
$y < 5$ Dividing both sides by 4.

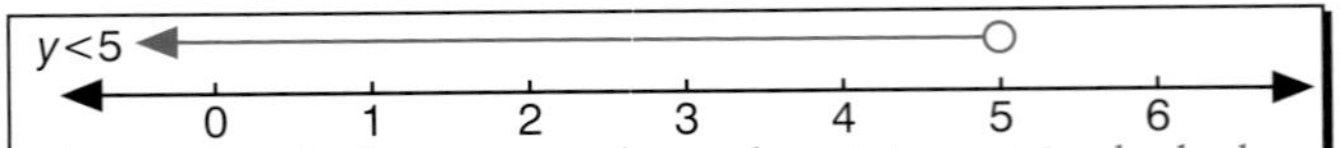

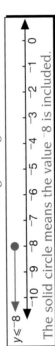

The open circle means the value 5 is not included.

e.g. $5 - \frac{1}{2}y \geqslant 9$
${}^-\frac{1}{2}y \geqslant 4$ Subtracting 5 from both sides.
$y \leqslant {}^-8$ Multiplying both sides by ${}^-2$ and reversing the sign.

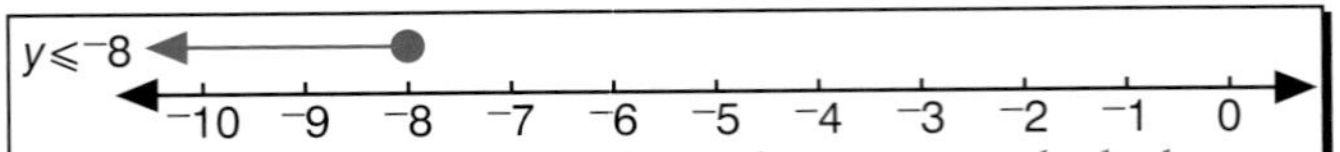

The solid circle means the value ${}^-8$ is included.

e.g. $7x < 8x + 5$
${}^-x < 5$ Subtracting $8x$ from both sides.
$x > {}^-5$ Multiplying both sides by ${}^-1$ and reversing the sign.

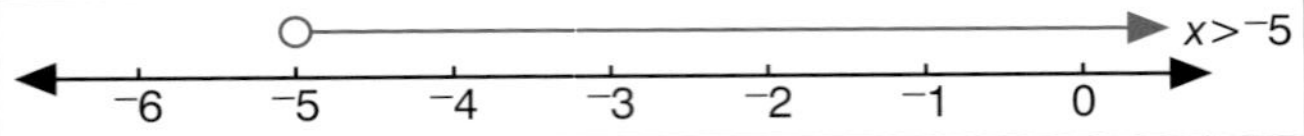

Patterns and sequences

A **sequence** is a set of numbers which follow a particular rule. The word 'term' is often used to describe the numbers in the sequence.

For the linear sequence

3, 7, 11, 15, 19, 23, …

the first term is 3 and the second term is 7 etc.

The expression 'the nth term' is often used to denote the value of any term in the sequence.

For the above sequence the nth term is $4n - 1$ so that:

the first term (where $n = 1$) is $4 \times 1 - 1 = 3$

the second term (where $n = 2$) is $4 \times 2 - 1 = 7$

the third term (where $n = 3$) is $4 \times 3 - 1 = 11$

Similarly:

the 50th term (where $n = 50$) is $4 \times 50 - 1 = 199$

and the 1000th term (where $n = 1000$) is $4 \times 1000 - 1 = 3999$

etc.

Sequence rules

Most **number sequences** involve adding/subtracting or multiplying/dividing according to some rule.

The following method can be used to find the nth term of **linear sequences**.

Linear sequences

For the sequence

3, 7, 11, 15, 19, 23, …

work out the differences as shown.

	3		7		11		15		19		23 …
1st difference		+4		+4		+4		+4		+4	…

As the first differences are all the same then the sequence is linear, so you can use the formula:

the nth term = first term + $(n - 1) \times$ 1st difference

For this sequence:

$$\begin{aligned}\text{the } n\text{th term} &= 3 + (n - 1) \times 4 \\ &= 3 + 4n - 4 \\ &= 4n - 1\end{aligned}$$

Special sequences

The following are special sequences of numbers that you should be able to recognise.

- 1, 4, 9, 16, 25, ... square numbers
- 1, 8, 27, 64, 125, ... cube numbers
- 1, 3, 6, 10, 15, ... triangle numbers
- 2, 3, 5, 7, 11, 13, 17, ... prime numbers

NB See Number cards 1 and 2 for more information on these special sequences.

NB You might find it helpful to know that the rule for the *n*th term of the triangle numbers is $\frac{1}{2}n(n + 1)$.

Fibonacci sequence

You should also be familiar with the Fibonacci sequence, where each term is found by adding the two previous terms.

1, 1, 2, 3, 5, 8, 13, 21, ...

Quadratic sequences

Quadratic sequences are more difficult. The rule for finding the *n*th term includes a term in n^2. You should be able to recognise some of the following quadratics.

		*n*th term
e.g.	1, 4, 9, 16, 25	n^2
	2, 5, 10, 17, 26	$n^2 + 1$
	0, 3, 8, 15, 24	$n^2 - 1$
	2, 8, 18, 32, 50	$2n^2$

But watch out for

	0, 1, 4, 9, 16	$(n - 1)^2$
and	4, 9, 16, 25, 36	$(n + 1)^2$

You should always check your answers to make sure they work for the given term.

Check yourself 6

Algebra 1–6

1 Find the sequence rule and write down the next three terms of each of the following sequences.

(a) $5, 2.5, 0, {}^{-}2.5, {}^{-}5, \ldots$

(b) ${}^{-}2, 4, {}^{-}8, 16, {}^{-}32, \ldots$

(1 *mark for each part*)

2 Simplify the following expressions.

(a) $3a - 5b - 2a + 5b$

(b) $x^3 + x^2 - 5x - 9x^2 + 7x$

(1 *mark for each part*)

3 Multiply out $3x(5 - 3x)$.

(1 *mark*)

4 Write down the *n*th term of this sequence.

$2, 7, 12, 17, \ldots$.

(3 *marks if fully correct*)

5 Expand and simplify these expressions.

(a) $(3x + 5)(2x - 7)$

(b) $(7x - 3)^2$

(1 *mark for each part*)

6 Given that $(2x - 1)(x + 3) = 2x^2 + ax + b$ find a and b.

(1 *mark for each part*)

7 Factorise $x^2 - 6x - 16$.

(2 *marks if fully correct*)

8 Find three consecutive numbers with sum of 72.

(1 *mark*)

9 Expand $(x + y)(x - y)$ and use this to work out $9999^2 - 9998^2$ without a calculator.

(2 *marks if fully correct*)

10 Find the value of the letters in the following equations.

(a) $3^x = 81$

(b) $5^{2x+1} = \frac{1}{125}$

(1 *mark for part* (*a*) *and* 2 *marks for part* (*b*))

Total marks = 20

ANSWERS & TUTORIALS

SCORE

1 **(a)** Rule: subtract 2.5; next three terms: $^-7.5, ^-10, ^-12.5$

(b) Rule: multiply by $^-2$; next three terms: $64, ^-128, 256$

2 **(a)** a The expression is $3a - 2a - 5b + 5b = a$.

(b) $x^3 - 8x^2 + 2x$ The expression is $x^3 + x^2 - 9x^2 - 5x + 7x = x^3 - 8x^2 + 2x$.

3 $15x - 9x^2$ $3x(5 - 3x) = 3x \times 5 + 3x \times {}^-3x$
$= 15x - 9x^2$

4 $5n - 3$

	2		7		12		17		…
1st difference		+5		+5		+5		…	

As the first differences are all the same the sequence is linear.
nth term = first term + $(n - 1) \times$ 1st difference
$= 2 + (n - 1) \times 5 = 2 + 5n - 5 = 5n - 3$

5 **(a)** $6x^2 - 11x - 35$
$(3x + 5)(2x - 7)$
$= 3x \times 2x + 3x \times {}^-7 + 5 \times 2x + 5 \times {}^-7$
$= 6x^2 - 21x + 10x - 35$
$= 6x^2 - 11x - 35$

(b) $49x^2 - 42x + 9$
$(7x - 3)^2$
$= (7x - 3)(7x - 3)$
$= 7x \times 7x + 7x \times {}^-3 - 3 \times 7x - 3 \times {}^-3$
$= 49x^2 - 21x - 21x + 9 = 49x^2 - 42x + 9$

SCORE

6 $a = {}^+5$ and $b = {}^-3$
$(2x - 1)(x + 3) = 2x^2 + 5x - 3$
so $a = {}^+5$ and $b = {}^-3$.

7 $(x + 2)(x - 8)$
$x^2 - 6x - 16 = (x \quad)(x \quad) = (x + 2)(x - 8)$
Looking for numbers that multiply together to give $^-16$.

8 23, 24 and 25
Let the consecutive numbers be x, $x + 1$ and $x + 2$.
Then $x + (x + 1) + (x + 2) = 72$
$3x + 3 = 72$
$3x = 69$
$x = 23$
So the numbers are 23, 24 and 25.

9 $x^2 - y^2$
$(x + y)(x - y) = x \times x + x \times {}^-y + y \times x + y \times {}^-y$
$= x^2 - y^2$

19 997
Using the fact that $x^2 - y^2 = (x + y)(x - y)$:
$9999^2 - 9998^2 = (9999 + 9998)(9999 - 9998)$
$= 19\,997$

10 **(a)** $x = 4$ $3^x = 81$ and $3^4 = 81$ so $x = 4$

(b) $x = {}^-2$ $5^{2x+1} = \frac{1}{125}$ and $5^{-3} = \frac{1}{125}$
so $2x + 1 = {}^-3$ and $x = {}^-2$

TOTAL SCORE OUT OF 20

Algebra 1–6

1 Write down the first five terms of a sequence where the nth term is given as:

(a) $5n - 7$

(b) $n^2 - 2n$

(1 *mark for each part*)

2 The formula for finding the angle sum of an n-sided polygon is $(2n - 4) \times 90°$.

Use the formula to find the angle sum of an 18-sided polygon.

(1 *mark*)

3 Expand and simplify the following expression.

$a(5a + 2b) - 2b(a - 5b^2)$

(2 *marks*)

4 Factorise these expressions completely.

(a) $2lw + 2wh + 2hl$

(b) $6x - 10x^2$

(1 *mark for each part*)

5 Write down the nth term of each sequence.

(a) 2, 5, 10, 17, 26, …

(b) 0, 1, 4, 9, 16, …

(1 *mark for each part*)

6 Factorise these expressions completely.

(a) $x^2 - 2x - 8$

(b) $2x^2 - 7x - 15$

(1 *mark for each part*)

7 Expand and simplify $(x + y)(y - z) - (x - y)(y + z)$.

(2 *marks*)

8 Simplify the expression $(x + 7)^2 - (x - 7)^2$.

(2 *marks*)

9 Three children are x years, $(x + 3)$ years and $2x$ years old. The sum of their ages is 27. How old is each of the children?

(2 *marks*)

10 Simplify the following.

(a) $(y^4)^2$

(b) $(2p^4)^5$

(c) $6x^6 \div 9x^8$

(1 *mark for each part*)

Total marks = 20

ANSWERS & TUTORIALS

1 (a) $^-2, 3, 8, 13, 18$

Substitute $n = 1, 2, 3, 4$ and 5 in $5n - 7$.

(b) $^-1, 0, 3, 8, 15$

Substitute $n = 1, 2, 3, 4$ and 5 in $n^2 - 2n$.

2 2880°

Angle sum $= (2 \times 18 - 4) \times 90° = 2880°$

3 $5a^2 + 10b^3$

$a(5a + 2b) - 2b(a - 5b^2)$
$= a \times 5a + a \times 2b - 2b \times a - 2b \times {}^-5b^2$
$= 5a^2 + 2ab - 2ab + 10b^3$
$= 5a^2 + 10b^3$

4 (a) $2(lw + wh + hl)$

As 2 is the only common factor.

(b) $2x(3 - 5x)$ As 2 and x are both factors.

5 (a) $n^2 + 1$

(b) $(n - 1)^2$

You should recognise the sequences as quadratic sequences. Compare with n^2.

(a)

	nth term
1, 4, 9, 10, 25	n^2
2, 5, 10, 17, 26	$n^2 + 1$

(b)

	nth term
1, 4, 9, 10, 25	n^2
0, 1, 4, 9, 16	$(n - 1)^2$

SCORE

6 (a) $(x + 2)(x - 4)$ **(b)** $(2x + 3)(x - 5)$

Compare with $(x \quad)(x \quad)$ and $(2x \quad)(x \quad)$, see examples on Algebra card 3.

7 $2y^2 - 2xz$

The expression is
$\{xy - xz + y^2 - yz\} - \{xy + xz - y^2 - yz\}$
$= xy - xz + y^2 - yz - xy - xz + y^2 + yz$
$= {}^-2xz + 2y^2$ or $2y^2 - 2xz$

8 $28x$

$a^2 - b^2 = (a + b)(a - b)$ so
$(x + 7)^2 - (x - 7)^2$
$= \{(x + 7) + (x - 7)\}\{(x + 7) - (x - 7)\}$
$= \{x + 7 + x - 7\}\{x + 7 - x + 7\}$
$= 2x \times 14$
$= 28x$

9 6, 9 and 12 years

Sum of ages is $x + (x + 3) + 2x = 4x + 3$
$4x + 3 = 27$ giving $x = 6$ and ages 6, 9 and 12 years.

10 (a) y^8 $(y^4)^2 = y^4 \times y^4 = y^{4+4} = y^8$

(b) $32p^{20}$ $(2p^4)^5 = 2p^4 \times 2p^4 \times 2p^4 \times 2p^4 \times 2p^4 = 32p^{20}$

(c) $\frac{2}{3x^2}$ $6x^6 \div 9x^8 = \frac{6x^6}{9x^8} = \frac{2}{3x^2}$

Cancelling down.

SCORE

TOTAL SCORE OUT OF 20

Linear graphs

Linear means 'straight line' so all linear graphs are straight lines and can be written in the form $y = mx + c$ where m is the **gradient** of the straight line and c is the **cut off** on the y-axis (also called the ***y*-intercept**).

Gradient of a straight line

The gradient of a line is defined as:

$$\frac{\text{vertical distance}}{\text{horizontal distance}}.$$

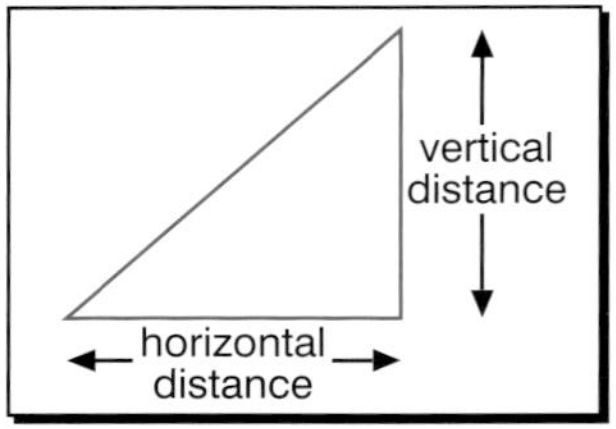

Gradients can be positive or negative depending on their direction of slope.

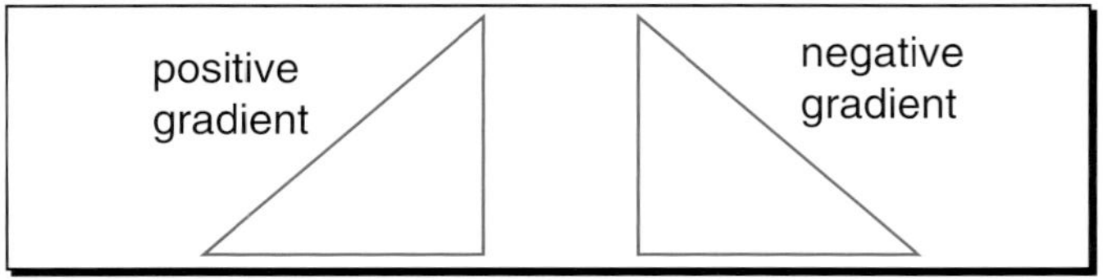

NB Parallel lines have the same gradient, and lines with the same gradient are parallel.

Quadratic and cubic graphs

Quadratic and **cubic graphs** all have the same basic shapes, as shown in the following sketches.

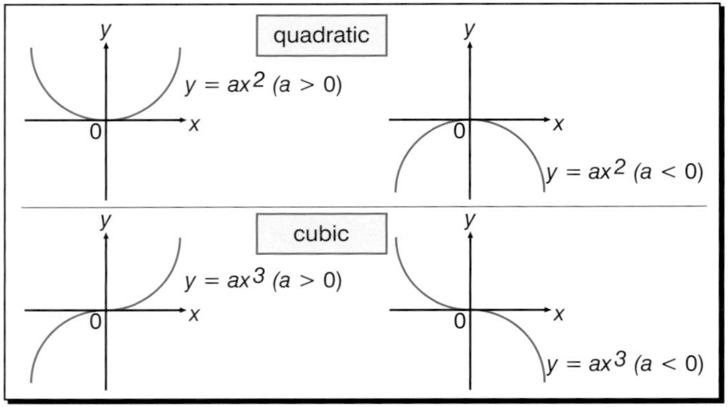

When drawing these graphs it is important to join the points with a smooth curve rather than a series of straight lines.

Reciprocal graphs

Reciprocal graphs are of the form $y = \frac{a}{x}$ and they all have the same basic shape, as illustrated here.

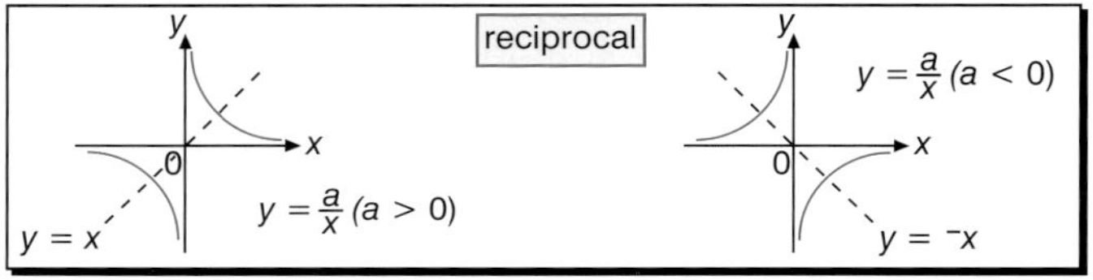

Graphing inequalities

Inequalities can easily be shown on a graph by replacing the inequality sign by an equals (=) sign and drawing this line on the graph. The two regions produced (one on either side of the line) can be defined using **inequality signs**.

It is usual to shade out the region which is not required, although some examination questions ask you to shade the required region. You must make it clear to the examiner which is your required region, by labelling it as appropriate. You also need to make it clear whether the line is **included** (i.e. the inequality is $\leqslant$ or $\geqslant$), or **excluded** (i.e. the inequality is $<$ or $>$).

e.g. Draw graphs of these lines.

$x = 2 \quad y = 1 \quad x + y = 6$

Use the graphs to identify and label the region where the points (x, y) satisfy the inequalities:

$x \geqslant 2 \quad y < 1 \quad x + y \leqslant 6$

NB The required region is indicated on the graph below and includes the adjoining parts of $x + 2$ and $y = 6 - x$ but not the line $y = 1$ (which is dotted to make this clear to the reader or examiner).

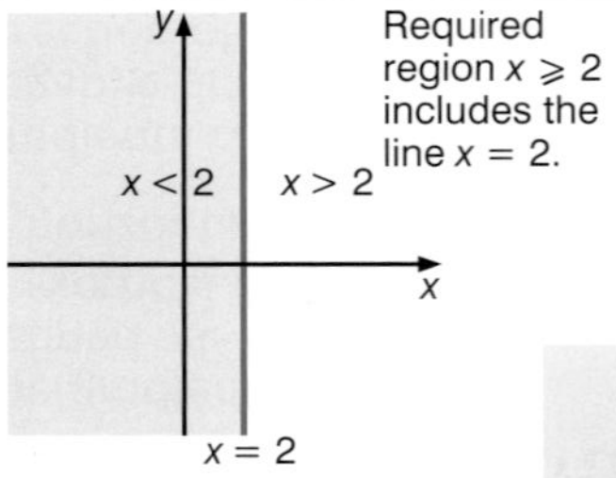

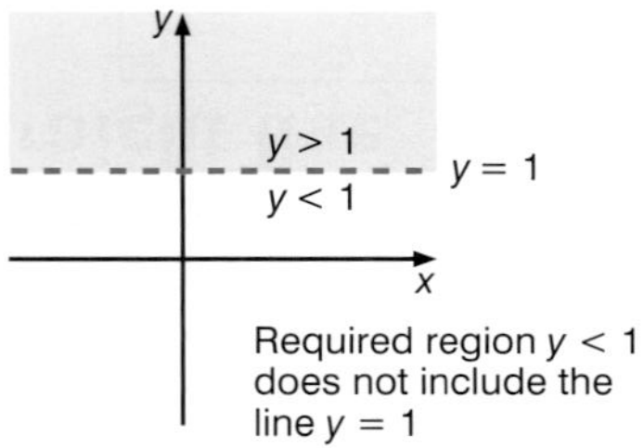

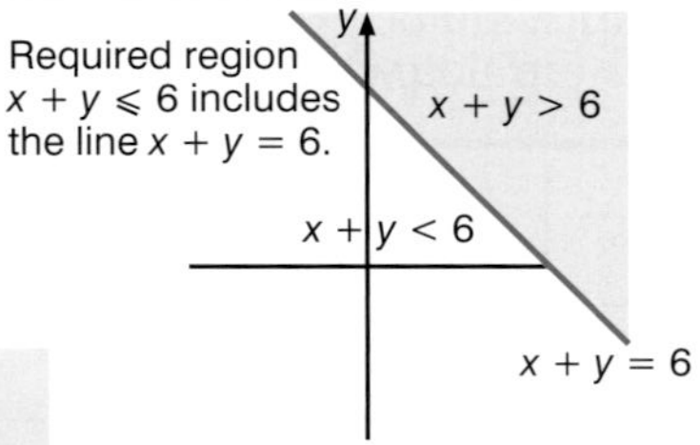

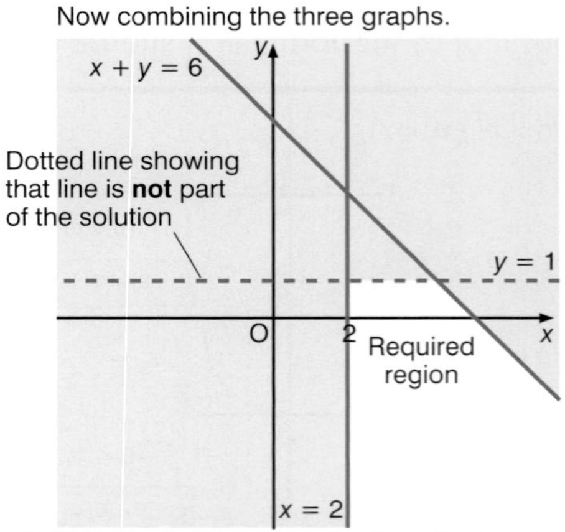

Quadratic equations

Quadratic equations are equations of the form $ax^2 + bx + c = 0$ where $a \neq 0$. Quadratic equations can be solved in a number of ways but at this level they are usually solved by graphical (see Algebra card 7) or algebraic methods.

Solution by factors

If the product of two numbers is zero then one or both of the numbers must be zero.

If $ab = 0$ then

either $a = 0$ or $b = 0$,

or both $a = 0$ and $b = 0$.

e.g. Solve the quadratic equation $(x - 5)(x + 3) = 0$.

Use the fact that since the product of the two brackets is zero then the expression inside one or both of them must be zero.

i.e. either $(x - 5) = 0$ which implies that $x = 5$

or $(x + 3) = 0$ which implies that $x = {}^-3$.

So the solutions of the equation $(x - 5)(x + 3) = 0$ are $x = 5$ and $x = {}^-3$.

e.g. Solve the quadratic equation $x^2 - 6x - 27 = 0$.

To solve the equation, factorise the left-hand side of the equation (see Algebra card 3) and then solve as before.

Factorising the left-hand side of the equation:

$$x^2 - 6x - 27 = (x \quad)(x \quad)$$

Now search for two numbers which multiply together to give ${}^-27$.

Try substituting:

${}^+1 \times {}^-27$ gives ${}^-27$: $(x + 1)(x - 27) = x^2 - 26x - 27$

${}^-1 \times {}^+27$ gives ${}^-27$: $(x - 1)(x + 27) = x^2 + 26x - 27$

${}^+9 \times {}^-3$ gives ${}^-27$: $(x + 9)(x - 3) = x^2 + 6x - 27$

${}^-9 \times {}^+3$ gives ${}^-27$: $(x - 9)(x + 3) = x^2 - 6x - 27$ ✓

The quadratic equation can be written $(x - 9)(x + 3) = 0$ and since the product of the two brackets is zero then the expression inside one or both of them must be zero.

i.e. either $(x - 9) = 0$ which implies that $x = 9$

or $(x + 3) = 0$ which implies that $x = {}^-3$.

So the solutions of the equation $x^2 - 6x - 27 = 0$ are $x = 9$ and $x = {}^-3$.

NB Always check your answers by substituting them into the original equation.

Trial and improvement methods

Trial and improvement can be used to provide successively better approximations to the solution of a problem.

NB An initial approximation can usually be obtained by drawing a graph of the function or else trying a few calculations (using whole numbers) in your head.

e.g. The length of a rectangle is 2 cm greater than the width and the area of the rectangle is 30 cm^2. Use trial and improvement to obtain an answer correct to the nearest millimetre.

Width	Length	Area	Comments
4	6	24	too small
5	7	35	too large
Width must lie between 4 and 5.			
4.5	6.5	29.25	too small
Width must lie between 4.5 and 5 (near to 4.5).			
4.7	6.7	31.49	too large
Width must lie between 4.5 and 4.7.			
4.6	6.6	30.36	too large
Width must lie between 4.5 and 4.6.			
4.55	6.55	29.8025	too small
Width must lie between 4.55 and 4.6.			

Since 4.55 and 4.6 are both equal to 4.6 (correct to the nearest millimetre) then you can stop and say that the solution is 4.6 (correct to the nearest millimetre).

e.g. Given that a solution of the equation $x^3 - 3x = 25$ lies between 3 and 4, use trial and improvement to obtain an answer correct to 1 decimal place.

When $x = 3$ $\quad x^3 - 3x = 3^3 - 3 \times 3 = 18$

When $x = 4$ $\quad x^3 - 3x = 4^3 - 3 \times 4 = 52$

Solution lies between 3 and 4 (closer to $x = 3$).

Try $x = 3.5$ $\quad x^3 - 3x = 3.5^3 - 3 \times 3.5 = 32.375$

Solution lies between 3 and 3.5 (closer to $x = 3.5$).

Try $x = 3.3$ $\quad x^3 - 3x = 3.3^3 - 3 \times 3.3 = 26.037$

Solution lies between 3 and 3.3 (closer to $x = 3.3$).

Try $x = 3.2$ $\quad x^3 - 3x = 3.2^3 - 3 \times 3.2 = 23.168$

Solution lies between 3.2 and 3.3.

Try $x = 3.25$ $\quad x^3 - 3x = 3.25^3 - 3 \times 3.25$
$= 24.578\,125$

Solution lies between 3.25 and 3.3.

Since 3.25 and 3.3 are both equal to 3.3 (correct to 1 decimal place) then you can stop and say that the solution is 3.3 (correct to 1 decimal place).

Rearranging formulae

You can rearrange (or **transpose**) a formula in exactly the same way as you solve an equation. However, to maintain the balance, you must make sure that whatever you do to one side of the formula you also do to the other side of the formula.

e.g. For $S = \frac{D}{T}$, S is the **subject** of the formula.

The formula can be rearranged to make D or T the subject.

$ST = D$ Multiplying both sides of the formula by T.

$D = ST$ Turning the formula round to make D the subject.

Now, using $D = ST$:

$\frac{D}{S} = T$ Dividing both sides of the formula by S.

$T = \frac{D}{S}$ Turning the formula around to make T the subject.

Simultaneous equations

Simultaneous equations are usually solved by graphical or algebraic methods. Both methods are described in the sections that follow.

Graphical solution of simultaneous equations

Simultaneous equations can be solved by using graphs to plot the two equations. The coordinates of the point of intersection give the solutions of the simultaneous equations.

e.g. Solve these simultaneous equations.

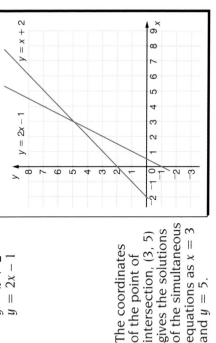

$y = x + 2$
$y = 2x - 1$

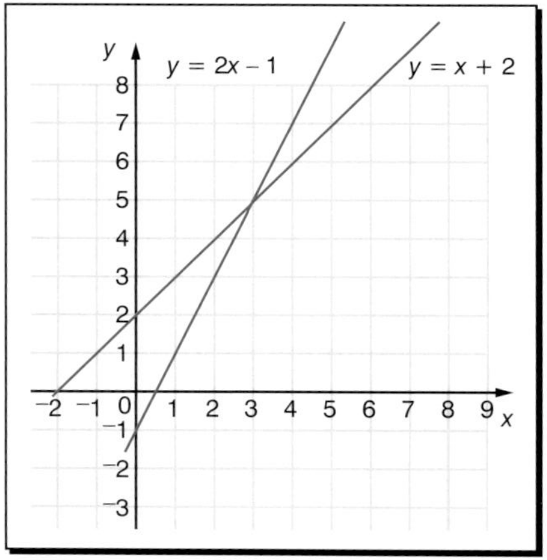

The coordinates of the point of intersection, (3, 5) gives the solutions of the simultaneous equations as $x = 3$ and $y = 5$.

NB Always check the solutions by substituting them into the original pair of equations.

Algebraic solution of simultaneous equations

The following two algebraic methods are commonly used to solve simultaneous equations.

Method of substitution

One equation is rewritten to make one of the unknowns the subject. This is then substituted into the second equation which can then be solved.

e.g. Solve these equations.

$$x - 4y = 11$$
$$3x + 4y = 1$$

Rewrite $x - 4y = 11$ as $x = 11 + 4y$.
Now substitute this value of x into the second equation.

$$\begin{aligned} 3x + 4y &= 1 \\ 3(11 + 4y) + 4y &= 1 \\ 33 + 12y + 4y &= 1 \\ 33 + 16y &= 1 \\ 16y &= {}^{-}32 \\ y &= {}^{-}2 \end{aligned}$$

Now substitute this value of y into $x = 11 + 4y$.

$x = 11 + 4 \times {}^{-}2$

$x = 3$

The solution is $x = 3$ and $y = {}^{-}2$, or $(3, {}^{-}2)$ as a coordinate pair.

Method of elimination

The equations, or multiples of the equations, are added or subtracted, to eliminate one of the unknowns. The resulting equation can then be solved.

e.g. Solve these equations.

$$x - 4y = 11$$
$$3x + 4y = 1$$

Add the left-hand sides of both equations together to eliminate y.
The sum of the two left-hand sides will equal the sum of the two right-hand sides.

$$\begin{aligned} (x - 4y) + (3x + 4y) &= 11 + 1 \\ x + 3x &= 12 \\ 4x &= 12 \\ x &= 3 \end{aligned}$$

Substituting this value in the first equation:

$$\begin{aligned} x - 4y &= 11 \\ 3 - 4y &= 11 \\ {}^{-}4y &= 8 \\ y &= {}^{-}2 \end{aligned}$$

The solution is $x = 3$, $y = {}^{-}2$, or $(3, {}^{-}2)$.

Both of these methods give the same solution of $x = 3$ and $y = {}^{-}2$.

Algebra 7–12

1 Rewrite the following with the letter indicated in brackets as the subject.

(a) $I = \frac{PRT}{100}$ (P) **(b)** $T = 2\pi\sqrt{\frac{l}{g}}$ (l)

(*1 mark each part*)

2 The graph shows the train journey from Asseema to Catalima and back again.

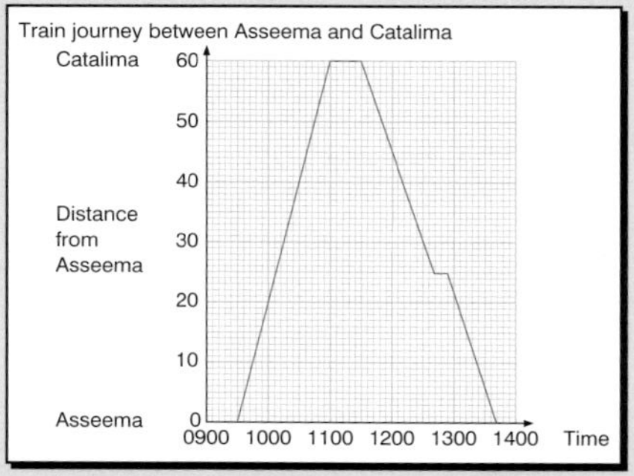

(a) What is the average speed from Asseema to Catalima?

The return journey includes a stop at Bereesa.

(b) How long does the train stop at Bereesa?

(c) What is the average speed between Bereesa and Asseema?

(*3 marks – deduct $\frac{1}{2}$ mark for each error*)

3 Calculate the gradients of the lines joining these pairs of points.

(a) (1, 1) and (6, 3) **(b)** (⁻3, 3) and (3, ⁻2)

(*1 mark each part*)

4 Write down the equations of the following graphs.

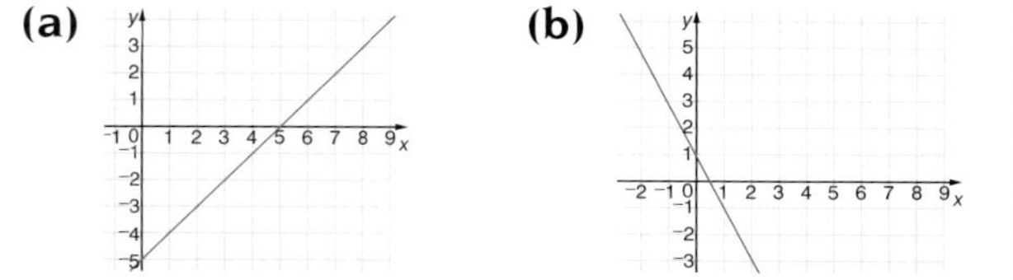

(*1 mark each part*)

5 List the integer values of $6 \leqslant 2x - 6 < 11$. (*1 mark*)

6 Show on a graph, the region represented by:
$x + y \leqslant 5$ $y < 2x$ $y \geqslant 0$ (*2 marks*)

7 By drawing suitable graphs, solve these simultaneous equations.
$x - y = 5$ $2x + y = 1$ (*2 marks*)

8 Draw the graph of $y = x^2 - 6x + 5$. Use it to find:

(a) the coordinates of the minimum value

(b) the values of x when $x^2 - 6x + 5 = 5$.

(*2 marks*)

9 Solve the following quadratic equations.

(a) $(x - 5)(x - 6) = 0$ **(b)** $x^2 + x - 12 = 0$

(*1 mark each part*)

10 A solution of the equation $x^3 + x = 100$ lies between 4 and 5. Use trial and improvement to find the solution correct to 1 d.p. (*2 marks*)

Total marks = 20

ANSWERS & TUTORIALS

SCORE

1 (a) $P = \dfrac{100I}{RT}$

$100I = PRT$ Multiplying both sides by 100 and swapping sides.

$P = \dfrac{100I}{RT}$ Dividing both sides by RT.

(b) $l = \left(\dfrac{T}{2\pi}\right)^2 g$

$\dfrac{T}{2\pi} = \sqrt{\dfrac{l}{g}}$ Dividing both sides by 2π.

$\left(\dfrac{T}{2\pi}\right)^2 = \dfrac{l}{g}$ Squaring both sides.

$l = \left(\dfrac{T}{2\pi}\right)^2 g$ Multiplying both sides by g.

2 (a) 40 kph Speed = 60 ÷ 1.5 = 40 kph

(b) 12 minutes

(c) 31.25 kph Speed = 25 ÷ (48 ÷ 60) = 31.25 kph

3 (a) $\frac{2}{5}$

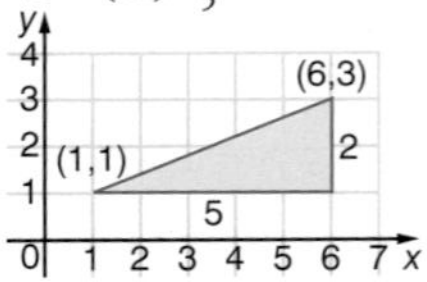

Gradient $= \frac{2}{5}$

(b) $-\frac{5}{6}$

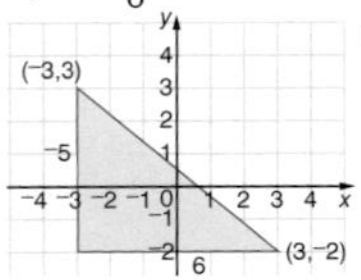

Gradient $= \frac{^{-}5}{6}$

4 (a) $y = x - 5$ From graph, $m = 1$, $c = {}^{-}5$ so $y = x - 5$.

(b) $y = {}^{-}2x + 1$ From graph, $m = {}^{-}2$, $c = 1$ so $y = {}^{-}2x + 1$.

SCORE

5 6, 7 and 8

$12 \leqslant 2x < 17$ Adding 6 to both sides .

$6 \leqslant x < 8\frac{1}{2}$ Dividing both sides by 2.

Integers satisfying the inequality are 6, 7 and 8.

6

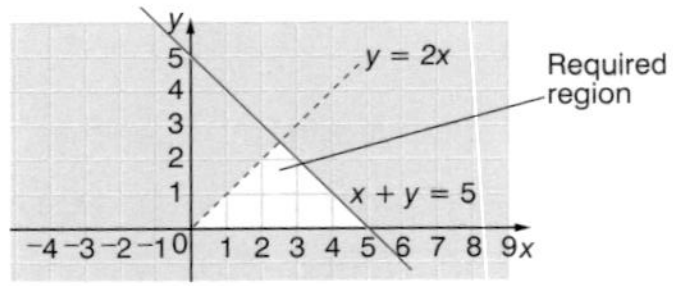

7 $x = 2$ and $y = {}^{-}3$

$y = x - 5$

So $m = 1$ and $c = {}^{-}5$.

$y = {}^{-}2x + 1$

So $m = {}^{-}2$ and $c = 1$.

The coordinates $(2, {}^{-}3)$ give the solution.

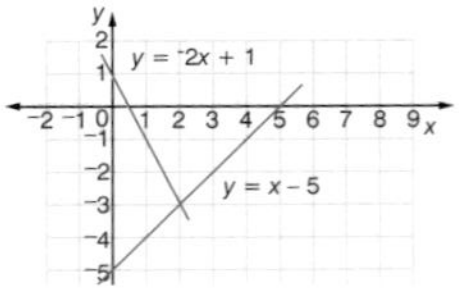

8 (a) $(3, {}^{-}4)$ Coordinates of minimum are $(3, {}^{-}4)$.

(b) $x = 0$ and $x = 6$

Values lie on $y = x^2 - 6x + 5$ and $y = 5$, i.e. $x = 0$ and $x = 6$.

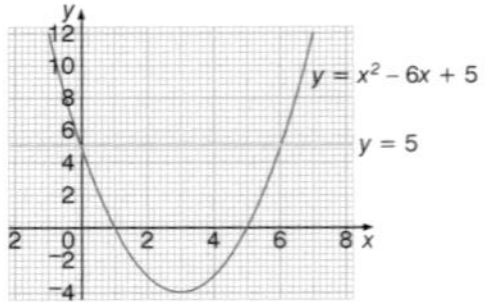

9 (a) $x = 5$ or $x = 6$ See Algebra card 9.

(b) $x = 3$ or $x = {}^{-}4$ See Algebra card 9.

10 4.6 (1 d.p.)

When $x = 4.5$ $x^3 + x = 95.625$ small

When $x = 4.6$ $x^3 + x = 101.936$ large

When $x = 4.55$ $x^3 + x = 98.746\,375$ small

TOTAL SCORE OUT OF 20

Algebra 7–12

1 On the same set of axes, sketch the following graphs.

(a) $y + 2x = 3$ **(b)** $x = 2y + 3$

(1 *mark each part*)

2 Draw individual graphs to show each of the following inequalities.

(a) $x \geqslant 3$ **(b)** $y < {}^{-}2$
(c) $x + y \leqslant 5$ **(d)** $y > x^2$

(1 *mark each part*)

3 Solve these inequalities.

(a) $3x + 7 \geqslant 10 - 2x$

(b) $4(1 - x) < 16 - 2x$

(1 *mark each part*)

4 By drawing the graphs of $y = \frac{1}{x}$ and $y = x^2 - 1$

on the same axes, solve the equation $\frac{1}{x} = x^2 - 1$.

(2 *marks*)

5 Solve these simultaneous equations:
$x - 2y = 7$
$3x + 2y = 5$

(a) by the method of substitution

(b) by the method of elimination.

(1 *mark each part*)

6 Calculate the coordinates of the point of intersection of the lines $3x + 4y = 1$ and $2x - 3y = 12$.

(2 *marks*)

7 Solve $x^2 - 5x = 6$.

(2 *marks*)

8 The length of the diagonal of a square is 50 cm. Form an algebraic expression and solve it to find the length of each side.

(2 *marks*)

9 Using the method of trial and improvement, solve the equation $x^3 + 2x = 35$ correct to 1 decimal place.

(2 *marks*)

Total marks = 20

ANSWERS & TUTORIALS

1 **(a)**, **(b)**

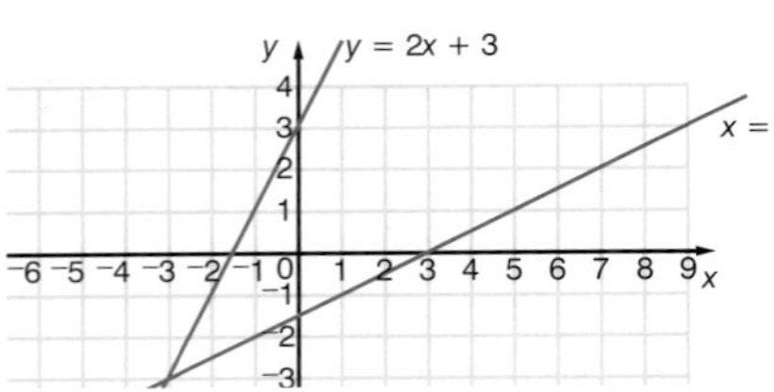

2

(a)

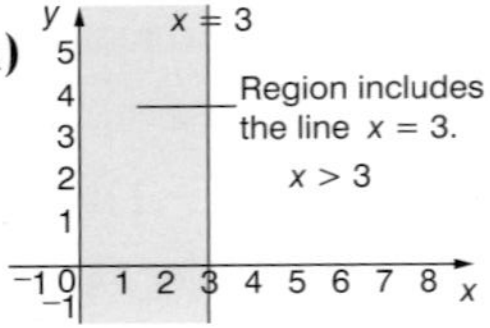

(b)

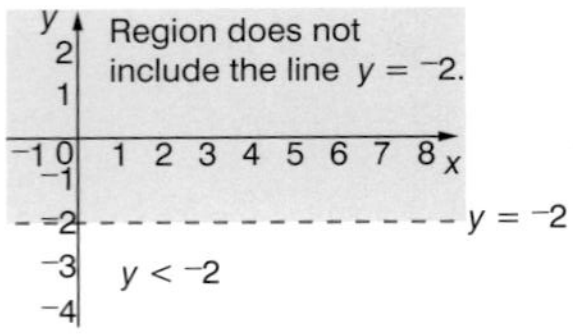

(c)

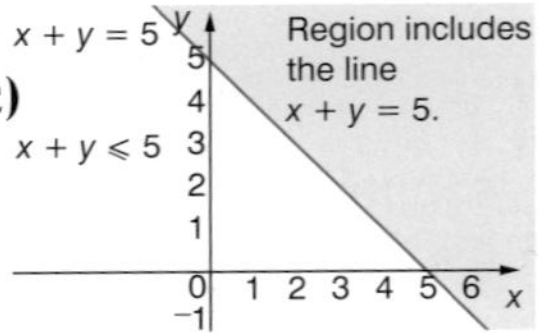

(d)

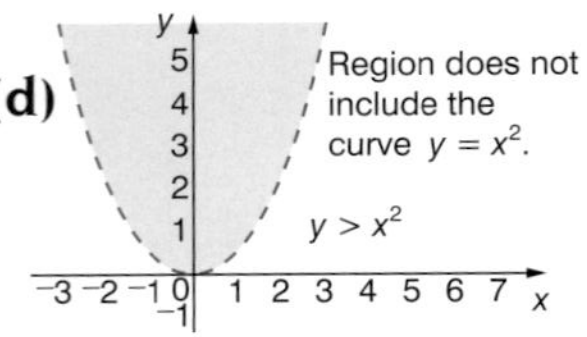

3 **(a)** $x \geqslant \frac{3}{5}$ See Algebra card 4.
(b) $x > {}^-6$ See Algebra card 4.

4 1.3 (1 d.p.)

From the graph, the curve $y = \frac{1}{x}$ crosses the curve $y = x^2 - 1$ when $x = 1.3$.

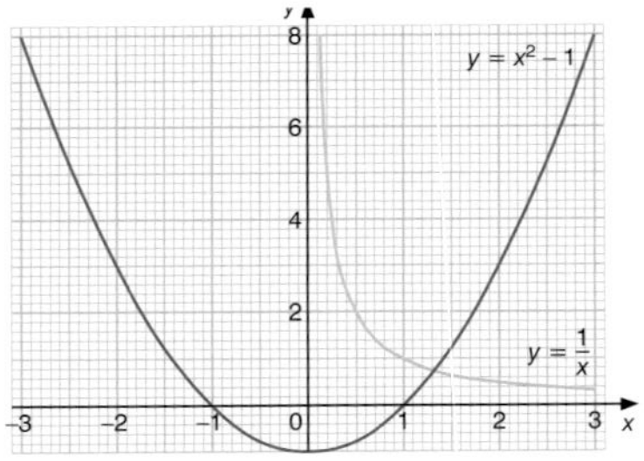

5 $x = 3$ and $y = {}^-2$

(a) If $x - 2y = 7$ then $x = 2y + 7$. Substitute this value for x.

(b) Adding the two equations, $4x = 12$, so $x = 3$.

6 $(3, {}^-2)$

$2 \times$ equation 1 gives $6x + 8y = 2$

$3 \times$ equation 2 gives $6x - 9y = 36$.

7 $x = 6$ or $x = {}^-1$ $\quad x^2 - 5x - 6 = (x - 6)(x + 1)$

8 35.4 cm (3 s.f.)

Using Pythagoras' theorem:

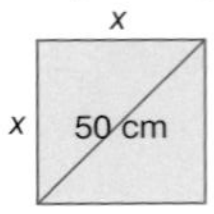

$x^2 + x^2 = 50^2$

$x = \pm 35.355\,339$

$x = 35.4$ cm (3 s.f.) ignoring negative length.

9 3.1 (1 d.p.)

First approximation (by graph or guessing) gives $x = 3$ and x lies between 3.05 and 3.1, giving $x = 3.1$ (1 d. p.). See page 42.

SCORE

SCORE

TOTAL SCORE OUT OF 20

Further quadratic equations

At the higher level you are expected to solve quadratic equations by a variety of methods, including the use of the formula and completing the square.

Solving quadratics by using the formula

Where a quadratic does not factorise then you can use the formula:

$$x = \frac{{}^{-}b \pm \sqrt{b^2 - 4ac}}{2a}$$

to solve any quadratic of the form $ax^2 + bx + c = 0$.

e.g. Solve the equation $x^2 + 2x - 1 = 0$.

First, compare $x^2 + 2x - 1 = 0$ with the general form $ax^2 + bx + c = 0$.

This gives $a = 1$, $b = 2$, $c = {}^{-}1$.

Substituting these values in the formula:

$$x = \frac{{}^{-}b \pm \sqrt{b^2 - 4ac}}{2a} = \frac{{}^{-}2 \pm \sqrt{2^2 - 4 \times 1 \times {}^{-}1}}{2 \times 1}$$

$$= \frac{{}^{-}2 \pm \sqrt{8}}{2}$$

So $x = \frac{{}^{-}2 + 2.828\,427\,1}{2} = 0.414$ (3 s.f.)

or $x = \frac{{}^{-}2 - 2.828\,427\,1}{2} = {}^{-}2.41$ (3 s.f.)

Your answers should be rounded to an appropriate degree of accuracy such as 3 s.f. or 2 d.p.

Alternatively, for the non-calculator paper, you should leave your answer in surd form, as follows.

$$x = \frac{{}^{-}2 \pm \sqrt{8}}{2}$$

$$x = \frac{{}^{-}2 + \sqrt{8}}{2} \text{ or } x = \frac{{}^{-}2 - \sqrt{8}}{2}$$

$$x = \frac{{}^{-}2 \pm 2\sqrt{2}}{2}$$ Simplifying further, as $\sqrt{8} = 2\sqrt{2}$

$$= {}^{-}1 \pm \sqrt{2}$$

So $x = {}^{-}1 + \sqrt{2}$

or $x = {}^{-}1 - \sqrt{2}$

Solving quadratics by completing the square

All quadratics can be rearranged into a square term and a constant term.

e.g. $x^2 + 2x + 1 = (x + 1)^2 + 0$

$x^2 + 2x + 2 = (x + 1)^2 + 1$

$x^2 + 2x + 3 = (x + 1)^2 + 2$ etc.

In general:

$x^2 + bx + c = (x + \frac{b}{2})^2 + \dots$ ← Adjust this term to make the two sides equal.

e.g. Solve the equation $x^2 + 2x - 1 = 0$ by completing the square.

$x^2 + 2x - 1 = 0$	Needs to be rewritten in the form $(x + 1)^2 + \dots$.
$\Rightarrow (x + 1)^2 - 2 = 0$	As this is the same as $x^2 + 2x - 1 = 0$.
$\Rightarrow (x + 1)^2 = 2$	Adding 2 to both sides.
$\Rightarrow (x + 1) = \pm\sqrt{2}$	Taking square roots on both sides and remember both roots.
$\Rightarrow x = {}^-1 \pm \sqrt{2}$	Subtracting 1 from both sides.

so $x = {}^-1 + \sqrt{2}$ or $x = {}^-1 - \sqrt{2}$ (as before, see Card 13.)

e.g. Solve the equation $2x^2 - 3x - 2 = 0$ by completing the square.

$2x^2 - 3x - 2 = 0$	
$\Rightarrow x^2 - \frac{3}{2}x - 1 = 0$	Dividing by 2 to get the coefficient of x^2 as unity.
$\Rightarrow (x - \frac{3}{4})^2 - \frac{25}{16} = 0$	$x^2 - \frac{3}{2}x - 1$ needs to be written in the form $(x - \frac{3}{4})^2 + \dots$.
$\Rightarrow (x - \frac{3}{4})^2 = \frac{25}{16}$	Adding $\frac{25}{16}$ to both sides.
$\Rightarrow x - \frac{3}{4} = \pm\sqrt{\frac{25}{16}}$	Taking square roots on both sides and remembering both roots.
$\Rightarrow x - \frac{3}{4} = \pm\frac{5}{4}$	Simplifying the right hand side by square rooting.
$\Rightarrow x = \frac{3}{4} \pm \frac{5}{4}$	Adding $\frac{3}{4}$ to both sides.
$\Rightarrow x = \frac{3}{4} + \frac{5}{4}$ or $x = \frac{3}{4} - \frac{5}{4}$	
$\Rightarrow x = 2$ or $x = {}^-\frac{1}{2}$	

NB It would have been much easier just to factorise the given equation.

Simplifying algebraic expressions

At the higher tier you will be expected to simplify harder algebraic expressions.

e.g. Simplify $\frac{x^2 - 4x}{x^2 - 16}$.

$$\frac{x^2 - 4x}{x^2 - 16}$$

$$= \frac{x(x - 4)}{(x - 4)(x + 4)}$$ Factorising the numerator. Factorising the denominator as the difference of two squares.

$$= \frac{x\cancel{(x - 4)}^{1}}{{}_{1}\cancel{(x - 4)}(x + 4)}$$ Cancelling $(x - 4)$ in the numerator and the denominator.

$$= \frac{x}{x + 4}$$

Further formula rearrangement

At the higher level you are required to rearrange formulae where the subject is found in more than one term.

e.g. Make b the subject of the formula $a = \frac{a - b}{ab}$.

$$a = \frac{a - b}{ab}$$

$$a^2b = a - b$$ Multiplying both sides by ab.

$$a^2b + b = a$$ Collecting the terms in b on one side.

$$b(a^2 + 1) = a$$ Factorising the left-hand side.

$$b = \frac{a}{a^2 + 1}$$ Dividing both sides by $(a^2 + 1)$.

Further simultaneous equations

Simultaneous equations at the higher tier may include a linear equation of the form $ax + by = c$ and an equation of the form $x^2 + y^2 = r^2$ or a linear equation plus a quadratic equation such as $y = ax^2 + bx + c$.

e.g. Solve these simultaneous equations.

$x^2 + y^2 = 100$
$x - y = 2$

Rewrite $x - y = 2$ as $x = y + 2$ and substitute the value of x into the first equation.

$x^2 + y^2 = 100$	
$(y + 2)^2 + y^2 = 100$	Where $x = y + 2$.
$y^2 + 4y + 4 + y^2 = 100$	Expanding.
$2y^2 + 4y + 4 = 100$	Simplifying.
$2y^2 + 4y - 96 = 0$	
$y^2 + 2y - 48 = 0$	
$(y + 8)(y - 6) = 0$	Factorising.
$y = {}^-8$ or $y = 6$	

Now substitute these values into $x = y + 2$.

When $y = {}^-8$, $x = {}^-6$.
When $y = 6$, $x = 8$.

So the solutions are $x = {}^-6$, $y = {}^-8$ and $x = 8$, $y = 6$.

e.g. Solve these simultaneous equations.

$y = x + 3$
$y = x^2 + 1$

Substitute $y = x + 3$ into the second equation.

$y = x^2 + 1$	
$x + 3 = x^2 + 1$	Where $y = x + 3$.
$0 = x^2 + 1 - x - 3$	Simplifying.
$0 = x^2 - x - 2$	
$x^2 - x - 2 = 0$	
$(x + 1)(x - 2) = 0$	Factorising.
$x = {}^-1$ or $x = 2$	

Now substitute these values into $y = x + 3$.

When $x = {}^-1$, $y = {}^-2$.
When $x = 2$, $x = 5$.

So the solutions are $x = {}^-1$, $y = 2$ and $x = 2$, $y = 5$.

NB The simultaneous equations can also be solved graphically by plotting $y = x + 3$ and $y = x^2 + 1$ and noting where the lines cross.

Further functions and graphs

There are four different graph transformations with which you need to be familiar.

$y = kf(x)$

$y = f(x) + a$

$y = f(kx)$

$y = f(x + a)$

Examples of each of these are shown for the function $f(x) = x^3$.

$y = kf(x)$

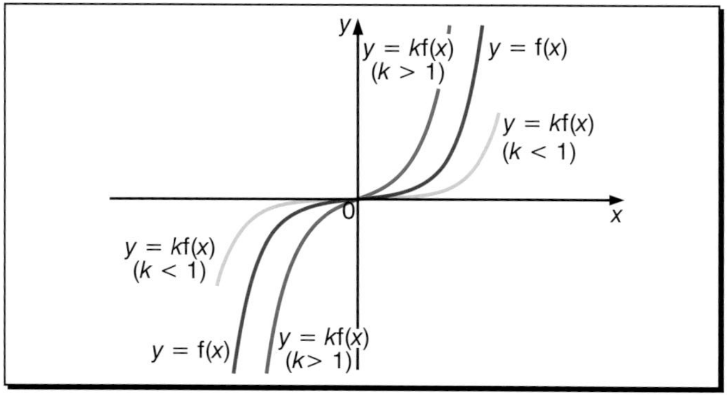

Under this transformation the graph of the function is stretched (or shrunk if $k < 1$) along the y-axis.

$y = f(x) + a$

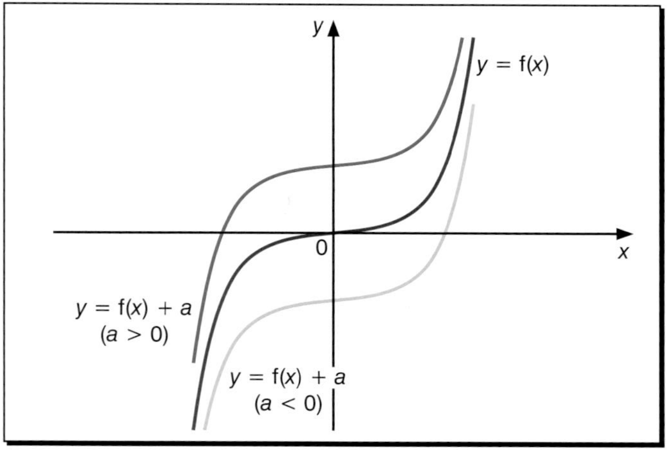

Under this transformation the function is translated along the y-axis. If $a > 0$ then the graph of the function moves up (positive direction) and if $a < 0$ then the graph of the function moves down (negative direction).

$y = f(kx)$

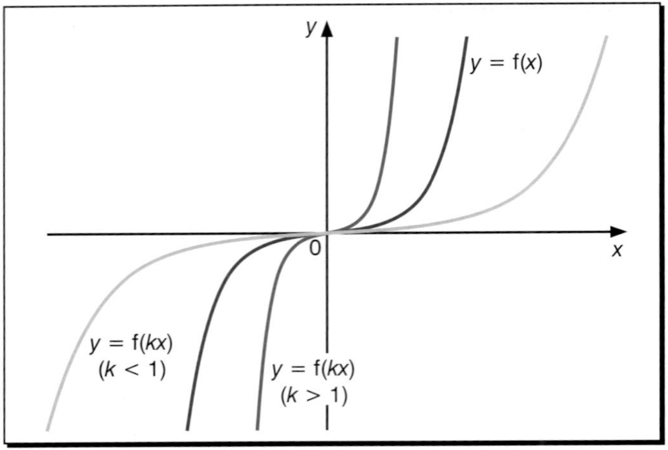

Under this transformation the graph of the function is shrunk (or stretched if $k < 1$) along the x-axis.

$y = f(x + a)$

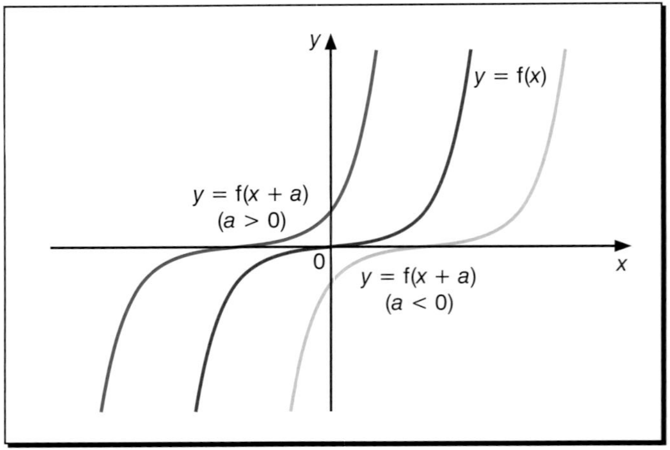

Under this transformation the function is translated along the x-axis. If $a > 0$ then the graph of the function moves to the left (negative direction) and if $a < 0$ then the graph of the function moves to the right (positive direction).

NB When using transformations it is helpful to try out a few points to check that you have the correct idea.

Algebra 13–18 (higher tier)

1 Solve the following quadratic formulae, giving your answers to 3 s.f.

(a) $x^2 + x - 1 = 0$

(b) $x^2 = 5x - 2$

(1 *mark each part*)

2 Solve the following quadratic formulae, leaving your answers in surd form.

(a) $x^2 + x - 1 = 0$

(b) $x^2 = 5x - 2$

(1 *mark each part*)

3 Rewrite $x^2 + 10x + 7$ in the form $(x + a)^2 + b$ where a and b are integers.

(2 *marks*)

4 Rearrange the formula $p = \frac{qr^2}{1 - r^2}$ to make r the subject.

(2 *marks if fully correct*)

5 Solve these simultaneous equations.

$y = 11x - 2$

$y = 5x^2$

(3 *marks if fully correct*)

6 A garden is 4 metres longer than it is wide. Find the width of the garden if the area is 50 square metres.

(2 *marks if fully correct*)

7 Sketch the following sets of graphs on graph paper, on the same set of axes.

(a) $y = x^2$ $y = x^2 + 3$ $y = (x - 2)^2$

(2 *marks for this part*)

(b) $y = \sin x$ $y = 3\sin x$ $y = \sin (x + 90°)$

(2 *marks for this part*)

8 A lorry travels 20 miles at an average speed of x miles per hour, then 45 miles at an average speed of $(x - 10)$ miles per hour. If the whole journey takes 2 hours, write down an equation in x and use it to find the value of x.

(3 *marks if fully correct*)

Total marks = 20

ANSWERS & TUTORIALS

1 **(a)** $x = 0.618$ or $x = {}^{-}1.62$ (3 s.f.)
See page 49.
(b) $x = 4.56$ or $x = 0.438$ (3 s.f.)
See page 49.

2 **(a)** $x = {}^{-}\frac{1}{2} \pm \frac{\sqrt{5}}{2}$
See page 49.
(b) $x = \frac{5}{2} \pm \frac{\sqrt{17}}{2}$
See page 49.

3 **(a)** $x^2 + 10x + 7 = (x + 5)^2 - 18$
So $a = 5$ and $b = {}^{-}18$

4 $r = \sqrt{\dfrac{p}{p + q}}$

$p(1 - r^2) = qr^2$	Multiplying by $(1 - r^2)$.
$p - pr^2 = qr^2$	Expanding the brackets.
$p = qr^2 + pr^2$	Collecting together terms in r^2.
$p = r^2(q + p)$	Factorising the RHS.
$\dfrac{p}{p + q} = r^2$	Dividing both sides by $p + q$.
$r = \sqrt{\dfrac{p}{p + q}}$	Taking square roots.

5 Solutions are $x = \frac{1}{5}$, $y = \frac{1}{5}$ or $x = 2$, $y = 20$
Substitute $y = 11x - 2$ into the second equation.

$y = 5x^2$
$11x - 2 = 5x^2$
$5x^2 - 11x + 2 = 0$
$(5x - 1)(x - 2) = 0$
$x = \frac{1}{5}$ or $x = 2$

When $x = \frac{1}{5}$, $y = \frac{1}{5}$ When $x = 2$, $y = 20$

6 5.35 metres (3 s.f.)
Let the width be w, then the length is $(w + 4)$ metres.
Area $= w(w + 4) = 50$ and $w^2 + 4w - 50 = 0$.
$w = 5.348\,469\,2$ or $w = {}^{-}9.348\,469\,2$ but a width of ${}^{-}9.348\,469\,2$ metres is impossible.
See page 49.

7 **(a)**

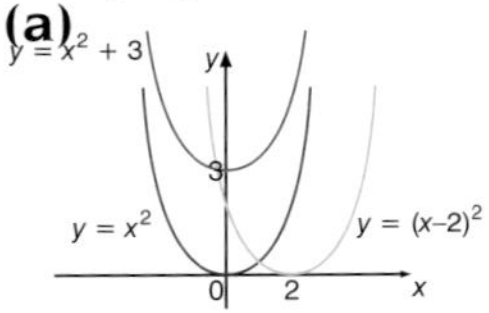

(b)

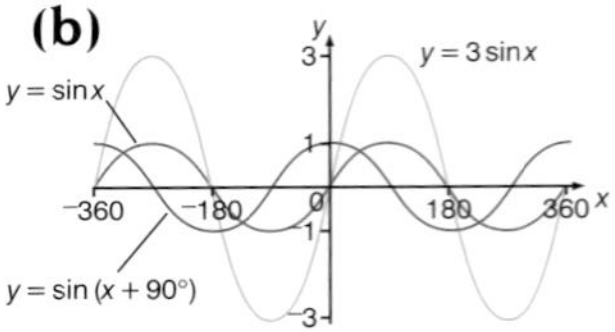

8 $x = 40$

Total time $= \dfrac{20}{x} + \dfrac{45}{x - 10} = 2$

$20(x - 10) + 45x = 2x(x - 10)$ Multiplying by $x(x - 10)$.
$2x^2 - 85x + 200 = 0$ Writing as a quadratic.
$(x - 40)(2x - 5) = 0$ Factorising.
$x = 40$ or $x = 2\frac{1}{2}$ but $x = 2\frac{1}{2}$ is not realistic.

TOTAL SCORE OUT OF 20

Angle properties

Angles on **a straight line** add up to 180°.

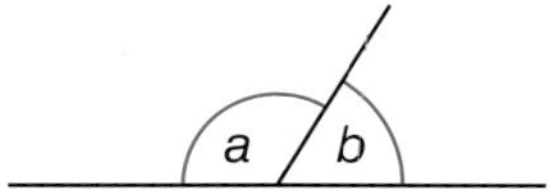

$a + b = 180°$

Angles **at a point** add up to 360°.

$p + q + r + s = 360°$

When two straight lines intersect the (**vertically**) **opposite** angles are equal.

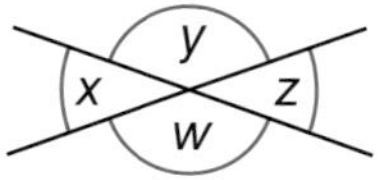

$x = z$ vertically opposite angles
$y = w$ vertically opposite angles

Angles between parallel lines

A **transversal** is a line which cuts two or more parallel lines.

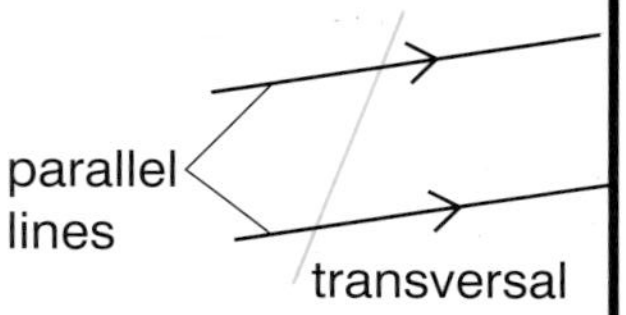

Corresponding angles (or F angles) are equal.

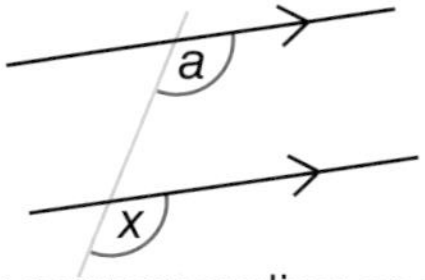

$a = x$ corresponding angles

Alternate angles (or Z angles) are equal.

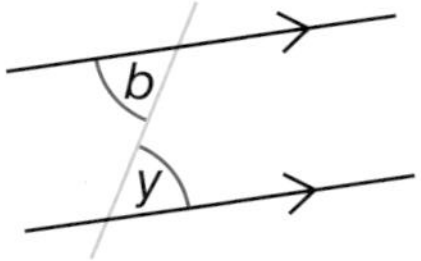

$b = y$ alternate angles

Interior angles add up to 180°.

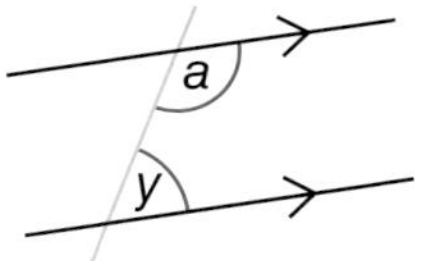

$a + y = 180°$ interior angles

e.g. Find the missing angles in this diagram.

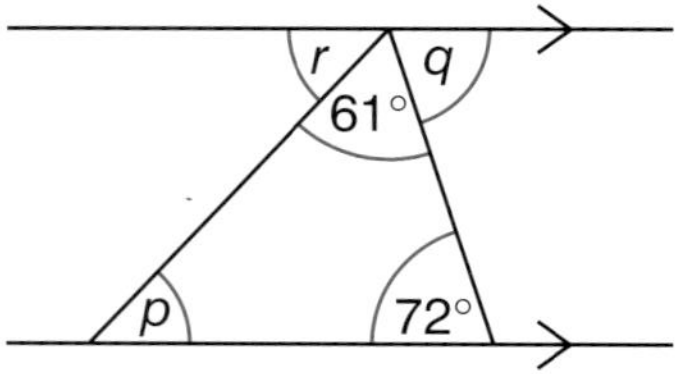

From the diagram:

$p = 47°$ The angles of the triangle add up to 180°.
$q = 72°$ It is an alternate angle between the two parallel lines.
$r = 47°$ Angles on a straight line add up to 180°, or r and p are alternate angles between two parallel lines.

NB In these examples there are usually several different ways of arriving at the same answer – trying different methods can serve as a useful check.

Bearings

Bearings are a useful way of describing directions. Bearings can be described in terms of the points on a compass or in terms of angles or turns measured from north *in a clockwise direction*. Bearings are usually given as three-figure numbers, so a bearing of 55° would usually be written as 055°.

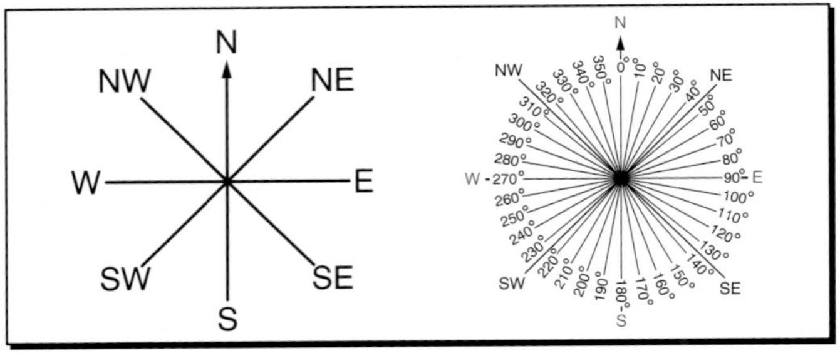

e.g. The bearing of a ship from a rock is 055°. What is the bearing of the rock from the ship?

From the sketch, it is clear that the required bearing is 235°.

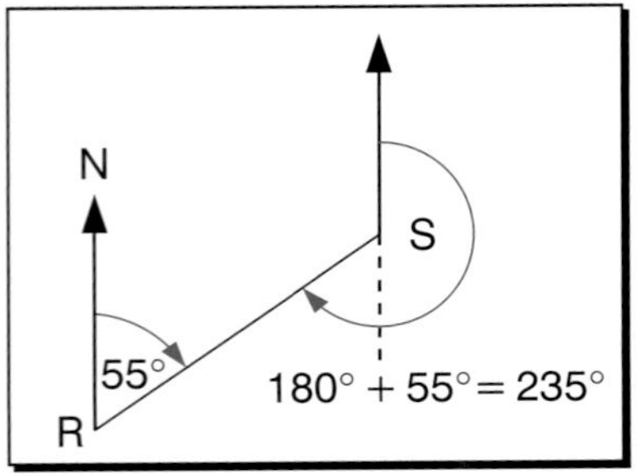

Parts of a circle

This diagram shows the main parts of a circle and the names given to them.

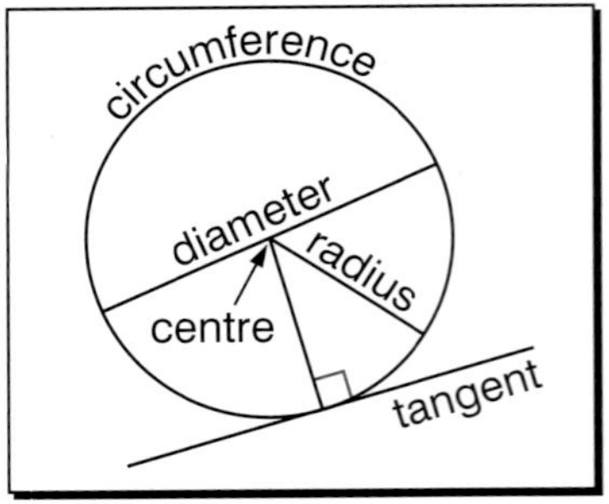

The following properties need to be known.

Circle properties

- The angle in a semicircle is always 90°.

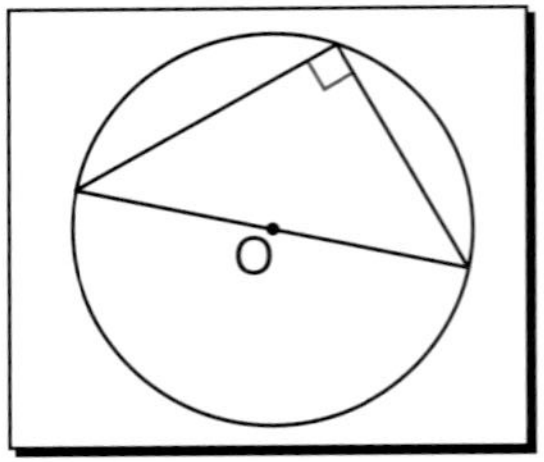

- The perpendicular bisector of a chord passes through the centre of the circle.

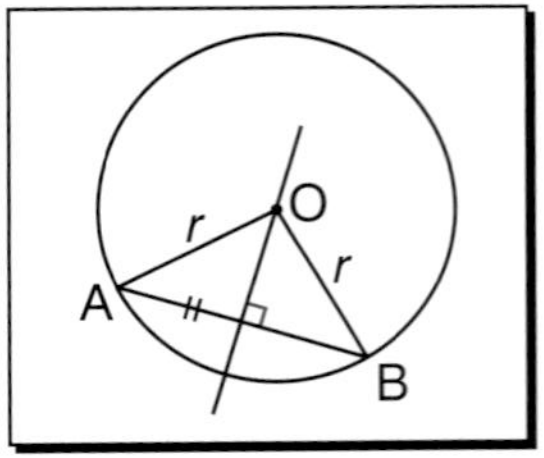

Angles properties in a circle

The following angle properties need to be known.

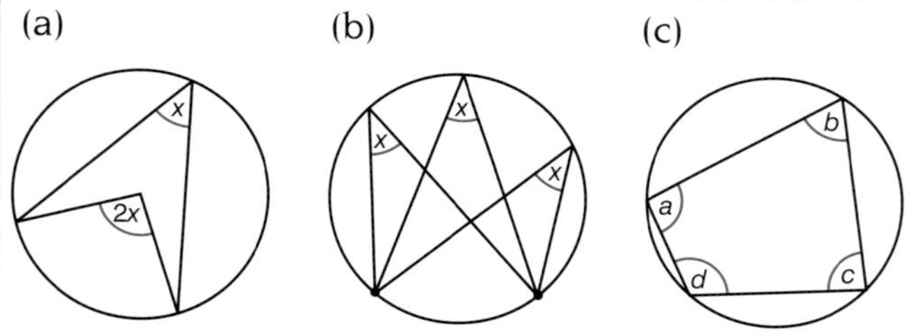

$a + c = 180°$
$b + d = 180°$

(a) The angle subtended by an arc (or chord) at the centre is twice that subtended at the circumference.

(b) Angles subtended by the same arc (or chord) are equal.

(c) The opposite angles of a cyclic quadrilateral are supplementary (i.e. they add up to 180°).

Tangents to a circle

A tangent is a straight line which touches a circle at only one point.

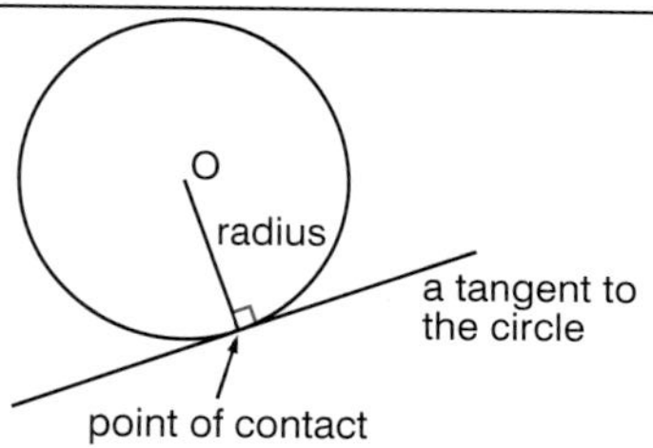

The following tangent properties need to be known.

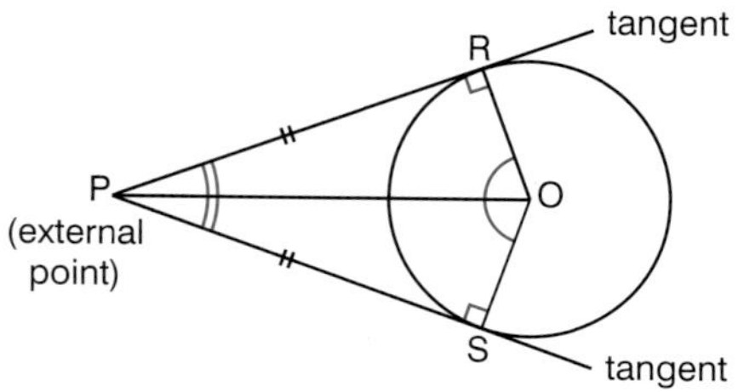

- A tangent to a circle is perpendicular to the radius at the point of contact.
- Tangents to a circle from an external point are equal in length.

Polygons

Any shape enclosed by straight lines is called a **polygon**. Polygons are named according to their number of sides.

Number of sides	Name of polygon
3	triangle
4	quadrilateral
5	pentagon
6	hexagon
7	heptagon or septagon
8	octagon
9	nonagon
10	decagon

A **regular** polygon has all sides equal and all angles equal.

A **convex** polygon has no interior angle greater than 180°.

A **concave** (or **re-entrant**) polygon has at least one interior angle greater than 180°.

Exterior angles

The exterior angle of a polygon is found by continuing the side of the polygon externally.

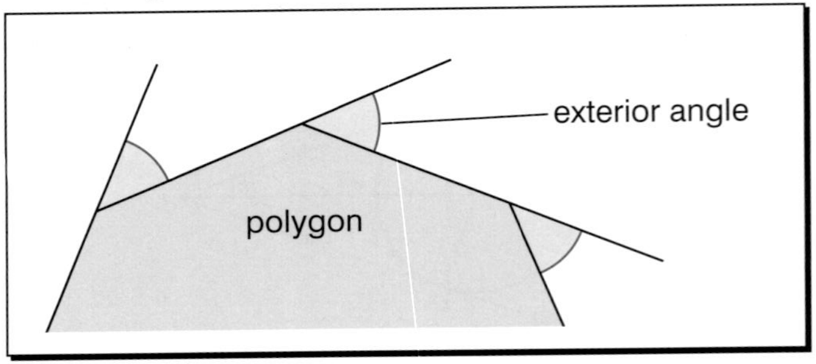

The exterior angle plus the adjacent interior angle add up to 180° (angles on a straight line).

The sum of all the exterior angles of a polygon is 360°.

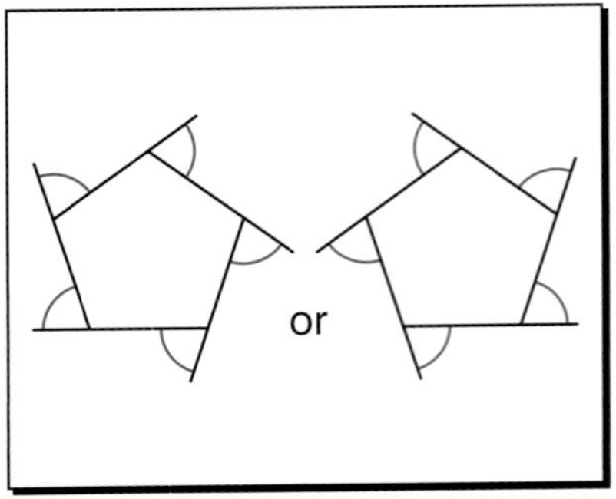

Angle sum of a polygon

The sum of the angles of any polygon can be found by dividing the polygon into triangles where each triangle has an angle sum of 180°.

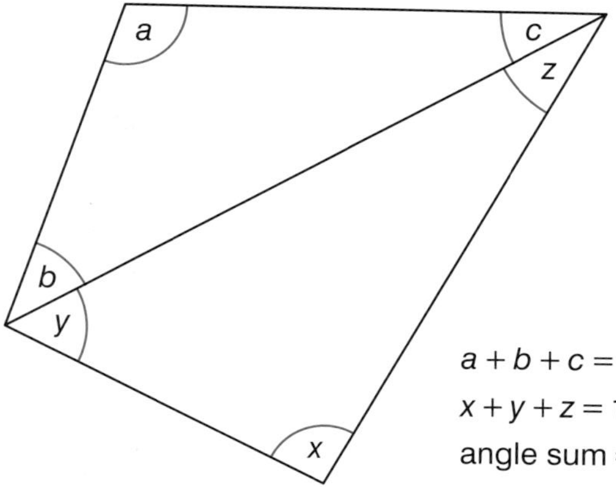

$a + b + c = 180°$

$x + y + z = 180°$

angle sum $= 360°$

A four-sided polygon can be split into two triangles.

Angle sum $= 2 \times 180°$
$= 360°$

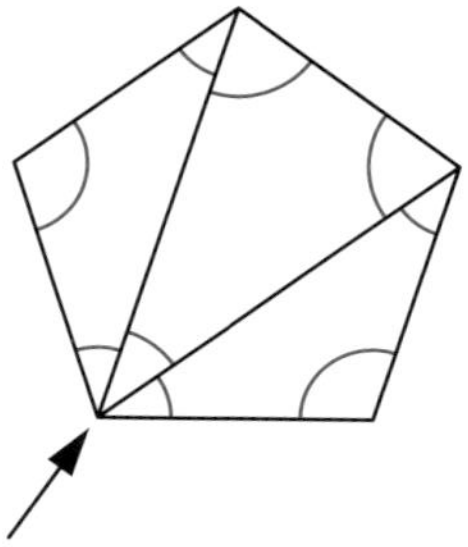

A five-sided polygon can be split into three triangles.

Angle sum $= 3 \times 180°$
$= 540°$

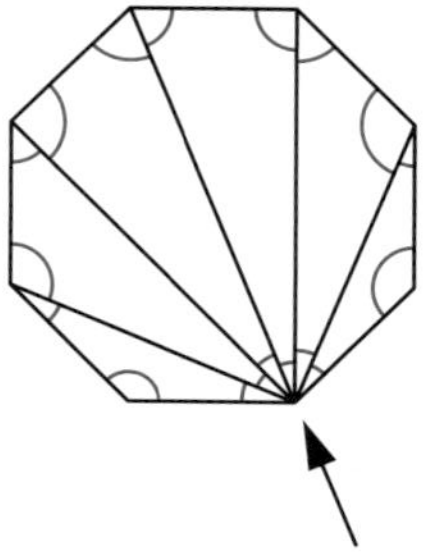

An eight-sided polygon can be split into six triangles.

Angle sum $= 6 \times 180°$
$= 900°$

From these diagrams, you can see that an n-sided polygon can be split into $(n - 2)$ triangles.

Angle sum of an n-sided polygon $= (n - 2) \times 180°$

NB The angle sum of an n-sided polygon can also be written as $(2n - 4) \times 90°$ or $(2n - 4)$ right angles.

Line symmetry

When a shape can be folded so that one half fits exactly over the other half, the shape is symmetrical and the fold line is called a **line of symmetry**.

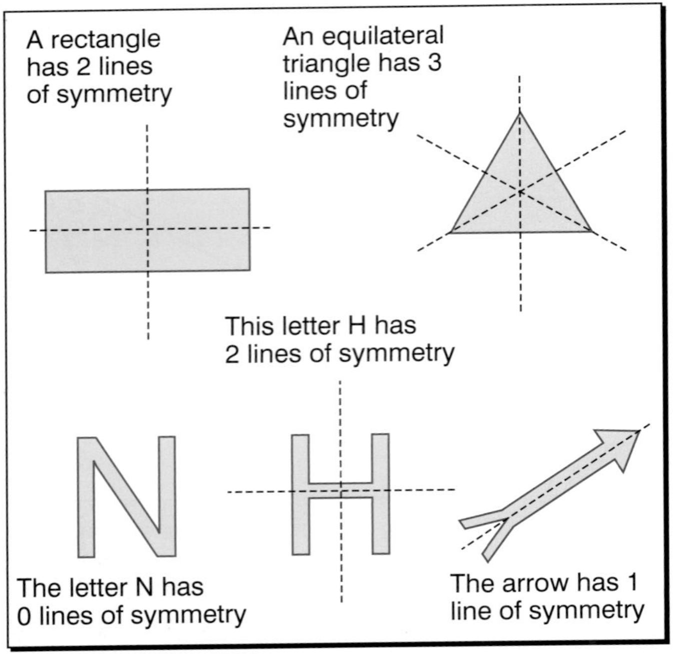
A rectangle has 2 lines of symmetry
An equilateral triangle has 3 lines of symmetry
This letter H has 2 lines of symmetry
The letter N has 0 lines of symmetry
The arrow has 1 line of symmetry

Rotational symmetry

When a shape can be rotated about its centre to fit exactly over its original position, the shape has **rotational symmetry**.

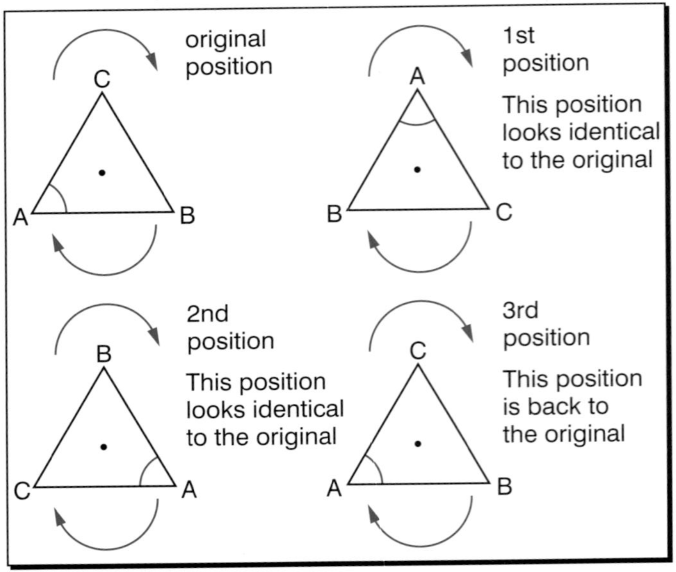
original position
A
B
C
1st position
This position looks identical to the original
2nd position
This position looks identical to the original
3rd position
This position is back to the original

The number of **different** positions gives the **order** of rotational symmetry. An equilateral triangle has rotational symmetry of order 3.

Planes of symmetry

A plane of symmetry divides a solid into two equal halves.

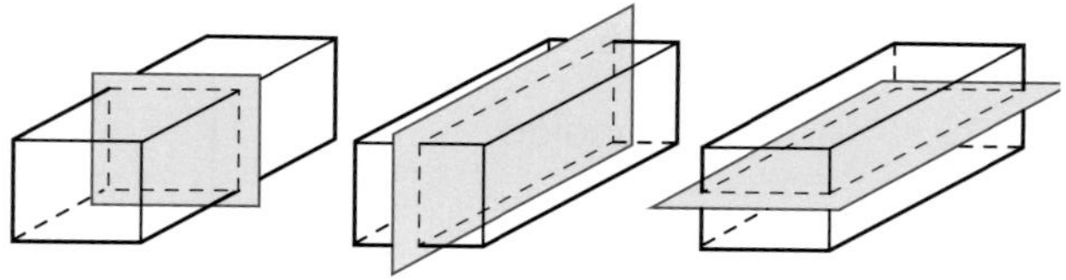

A cuboid has 3 planes of symmetry

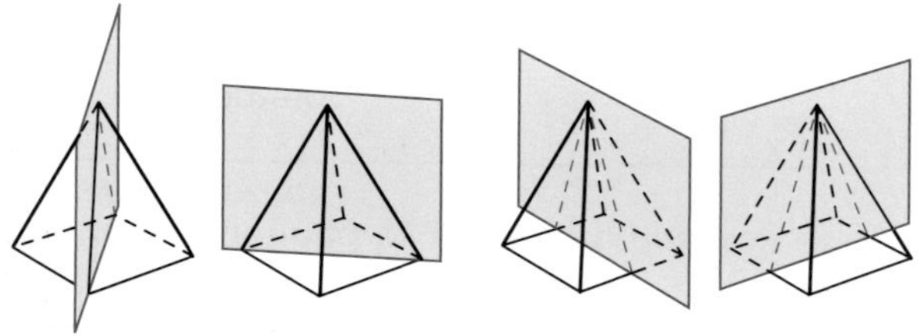

A square-based pyramid has 4 planes of symmetry

Tessellations

If congruent shapes fit together exactly to cover an area completely, then the shapes **tessellate**. These shapes tessellate.

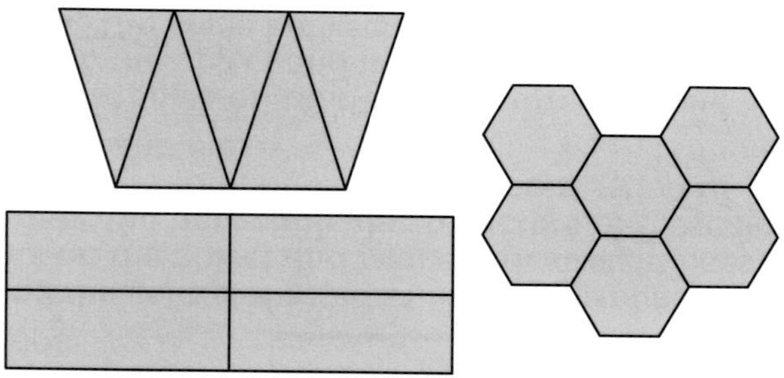

These are examples of shapes that do not tessellate.

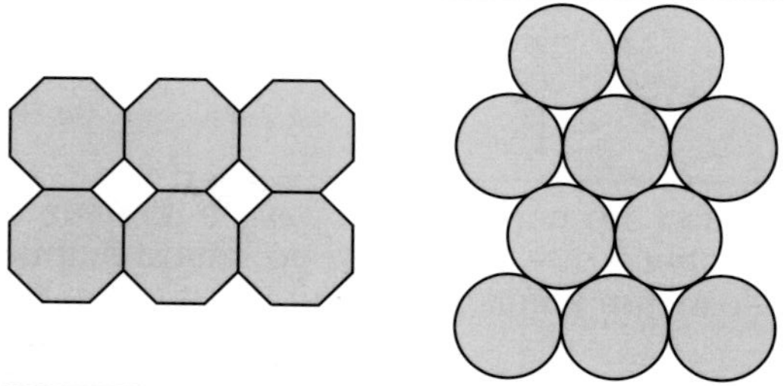

Reflection

A **reflection** is a **transformation** in which any two corresponding points on the object and image are the same distance away from a fixed line (called the **line of symmetry** or **mirror line**).

A reflection is defined by giving the position of the line of symmetry.

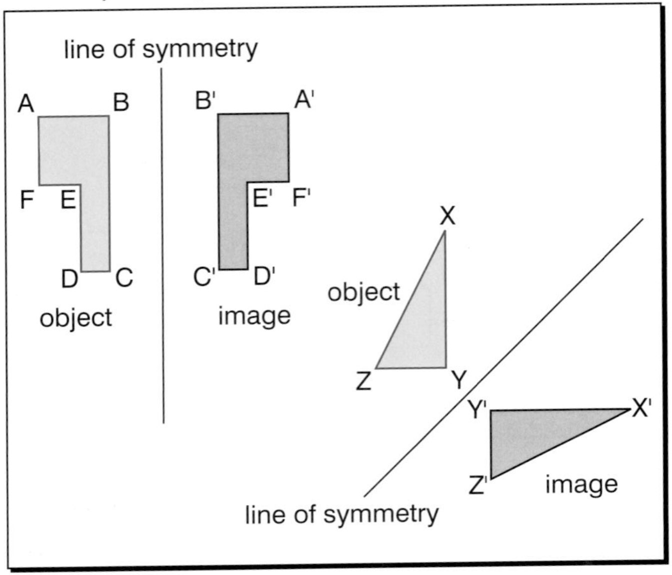

Rotation

A **rotation** is a transformation in which any two corresponding points on the object and image make the same angle at a fixed point (called the **centre of rotation**).

A rotation is defined by giving the position of the centre of rotation, along with the angle and the direction of the rotation.

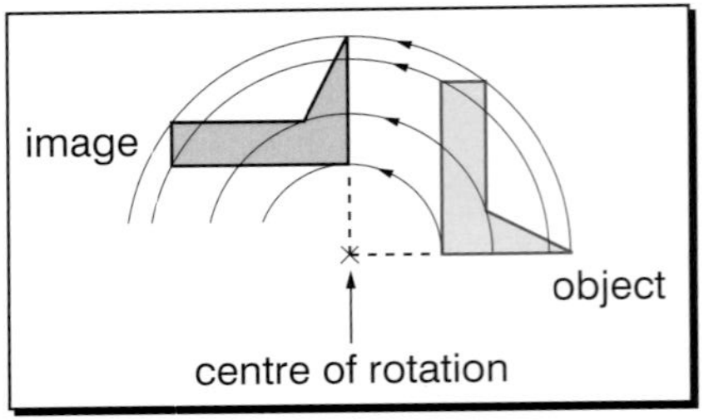

To find the centre of rotation, join corresponding points on the object and image with straight lines and draw the **perpendicular bisectors** of these lines. The centre of rotation lies on the intersection of these straight lines.

To find the angle of rotation, join corresponding points on the object and image to the centre of rotation. The angle between these lines is the **angle of rotation**.

NB In mathematics an anticlockwise turn is described as **positive** (+ve) and a clockwise turn is described as **negative** (–ve).

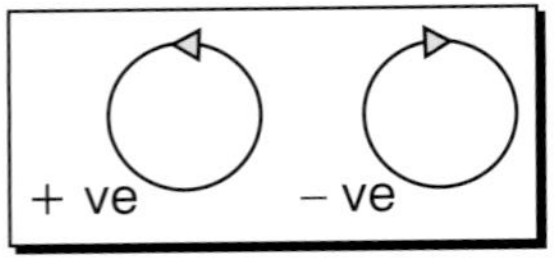

Enlargement

An **enlargement** is a transformation in which the distance between a point on the image and a fixed point (called the **centre of enlargement**) is a factor of the distance between the corresponding point on the object and the fixed point.

An enlargement is defined by giving the position of the centre of enlargement along with the factor (called the **scale factor**).

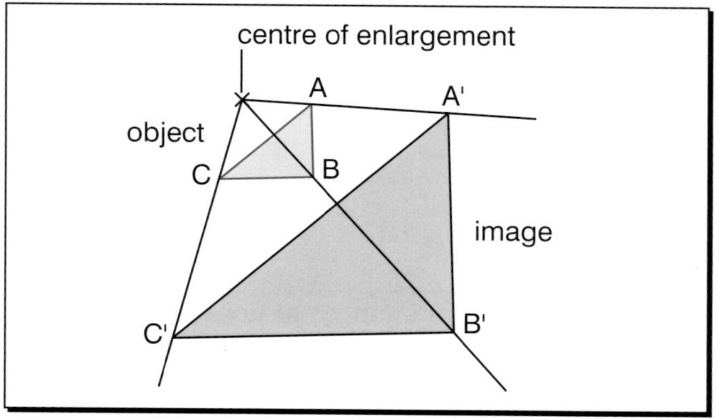

To find the centre of enlargement, join corresponding points on the object and image with straight lines. The centre of enlargement lies on the intersection of these straight lines.

The scale factor (SF) of an enlargement can be found as follows.

$$SF = \frac{\text{distance of point on image from centre}}{\text{distance of corresponding point on object from centre}}$$

or

$$SF = \frac{\text{distance between two points on image}}{\text{distance between corresponding points on object}}$$

e.g. The points A(3, 8), B(7, 8), C(7, $^-4$) and D(3, $^-2$) are joined to form a trapezium which is enlarged, scale factor $\frac{1}{2}$, with ($^-5$, $^-6$) as the centre.

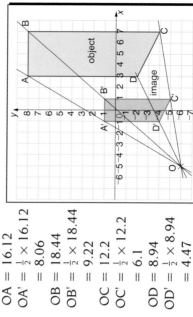

$OA = 16.12$
$OA' = \frac{1}{2} \times 16.12$
$= 8.06$
$OB = 18.44$
$OB' = \frac{1}{2} \times 18.44$
$= 9.22$
$OC = 12.2$
$OC' = \frac{1}{2} \times 12.2$
$= 6.1$
$OD = 8.94$
$OD' = \frac{1}{2} \times 8.94$
$= 4.47$

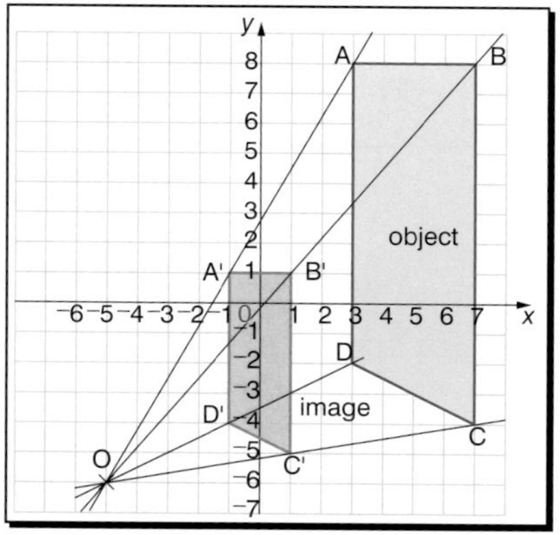

NB A fractional scale factor will reduce the object.

Translation

A **translation** is a transformation in which the distance and direction between any two corresponding points on the object and image are the same.

A translation is defined by giving the distance and direction of the translation.

This is a translation of six units to the right.

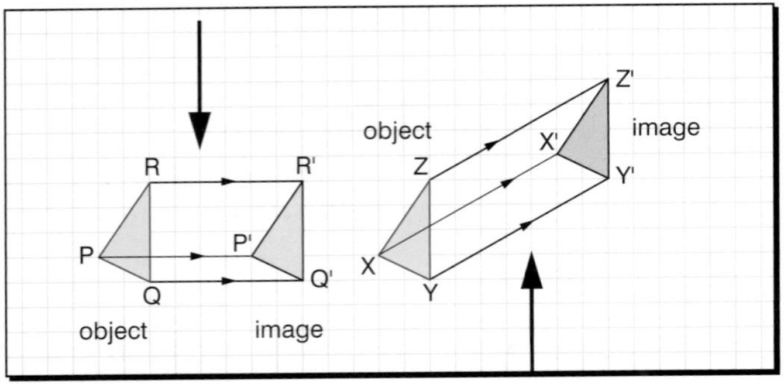

This is a translation of seven units right and four units up.

Vector notation is often used to describe translations. The notation is written as

$$\begin{pmatrix} \text{units moved in } x\text{-direction} \\ \text{units moved in } y\text{-direction} \end{pmatrix}$$

so the translations above can be written as $\begin{pmatrix} 6 \\ 0 \end{pmatrix}$ and $\begin{pmatrix} 7 \\ 4 \end{pmatrix}$.

e.g. The triangle ABC with coordinates (1, 1), (3, 2) and (2, 5) undergoes a translation of $\begin{pmatrix} 2 \\ ^-6 \end{pmatrix}$ to A'B'C'. Show the triangles ABC and A'B'C' and write down the translation that will return A'B'C' to ABC.

The vector $\begin{pmatrix} 2 \\ ^-6 \end{pmatrix}$ is a movement of 2 units to the right and $^-6$ units upwards (i.e. 6 units downwards).

The translation which will return A'B'C' to ABC is $\begin{pmatrix} ^-2 \\ 6 \end{pmatrix}$.

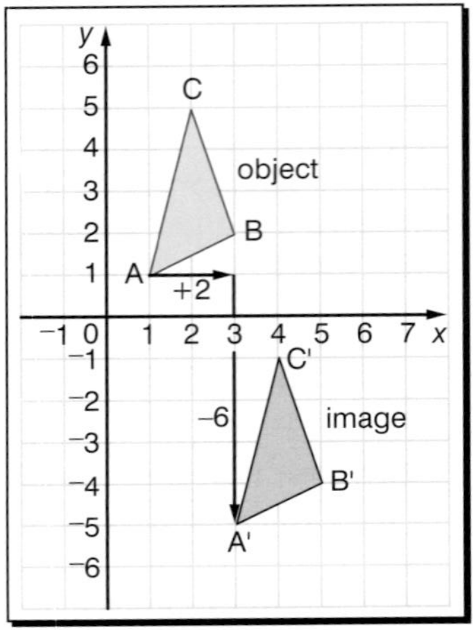

Combinations of transformations

A combination of the same transformation or a combination of different transformations can sometimes be described as a single transformation.

e.g. If R is a reflection in the y-axis and T is a rotation of $^-90°$ about the origin, show (on separate diagrams) the image of the triangle XYZ with vertices X(2, 1), Y(2, 5) and Z(4, 2) under these combined transformations:

(a) T followed by R

(b) R followed by T.

Which single transformation will return each of these combined transformations back to its original position?

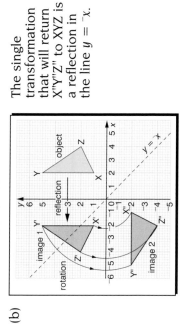

(a)

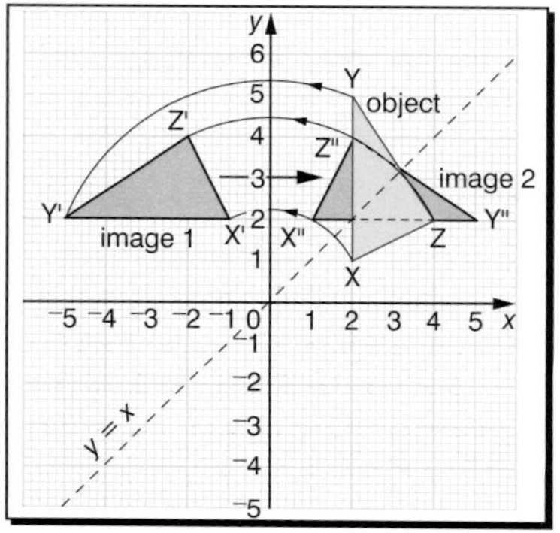

The single transformation that will return X'Y'Z' to XYZ is a reflection in the line $y = x$.

(b)

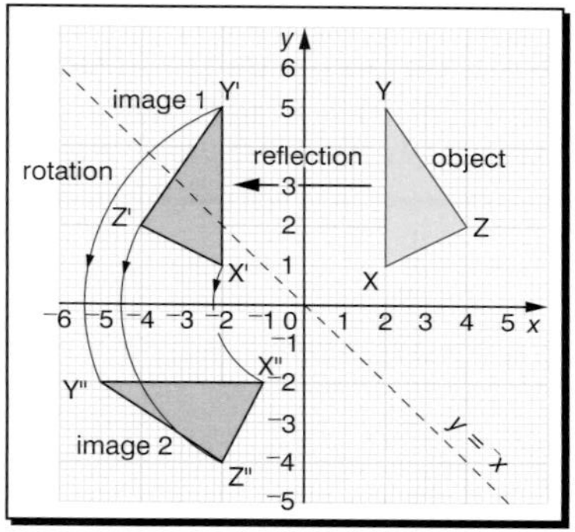

The single transformation that will return X"Y"Z" to XYZ is a reflection in the line $y = {}^-x$.

NB You can see from the example above that the order of the transformations is important.

Defining transformations

Reflection: define the line of reflection.

Rotation: define the centre of rotation and the angle of rotation (positive or negative).

Enlargement: define the centre of enlargement and the scale factor.

Translation: define the distance and the direction. You can also use vector notation.

Shape, space & measures 1–11

1 Use the diagram to calculate the sizes of the angles marked *a*, *b*, *c* and *d*.

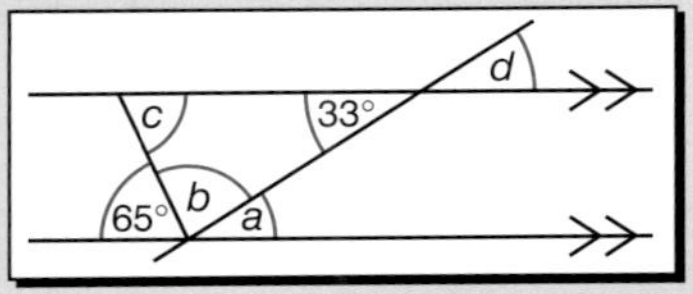

(1 *mark for each angle*)

2 **(a)** How many lines of symmetry do the following shapes have?
(i) kite ($\frac{1}{2}$ *mark*)
(ii) regular hexagon ($\frac{1}{2}$ *mark*)

(b) How many planes of symmetry do the following solids have?
(i) equilateral triangular prism ($\frac{1}{2}$ *mark*)
(ii) square-based pyramid ($\frac{1}{2}$ *mark*)

3 A plane takes off heading north-east and then makes a left-hand turn through 90°. What bearing is the plane now headed on?
(2 *marks*)

4 Find the angle sum of a heptagon. (2 *marks*)

5 The points A, B and C lie on a circle so that AB passes through the centre, O, and ∠CAB = 35°. Calculate ∠ACO, ∠ACB and ∠OCB.
(1 *mark for each angle*)

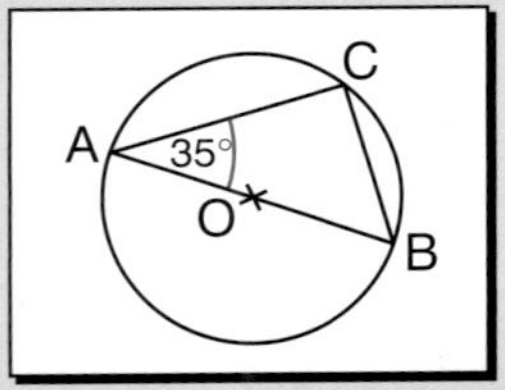

6 Chords AC and BD are perpendicular. ∠DAC = 62°.

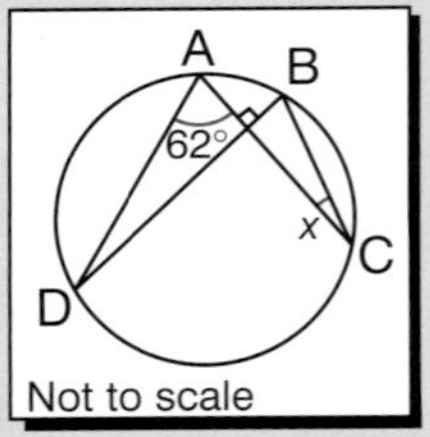

(a) Calculate the value of *x*. (2 *marks*)

(b) P is a point on the minor arc CD. Explain why ∠DPC is 118°. (1 *mark*)

7 An enlargement scale factor $\frac{1}{4}$ and centre (0, 0) transforms parallelogram ABCD onto $A_1B_1C_1D_1$. Then parallelogram $A_1B_1C_1D_1$ is translated by vector $\begin{pmatrix} 4 \\ -2 \end{pmatrix}$ onto $A_2B_2C_2D_2$. What are the coordinates of the points A_2, B_2, C_2 and D_2?

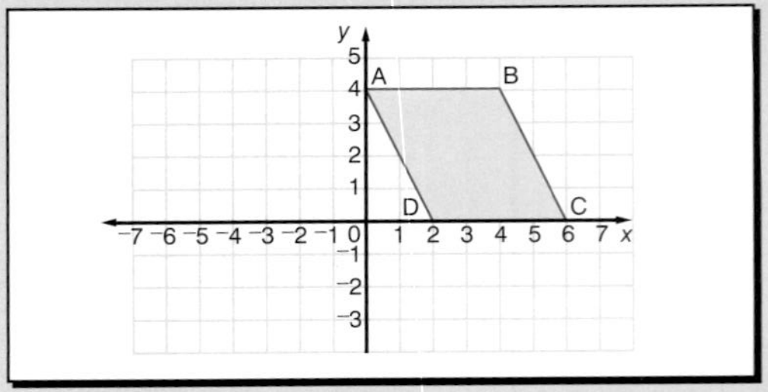

(2 *marks*)

8 The triangle PQR with vertices P(2, 1), Q(3, 1) and R(2, 3) is rotated through ⁺90° about the origin, then reflected in the line $y = x$. What single transformation will map the triangle back to its original position? (2 *marks*)

***Total marks* = 20**

ANSWERS & TUTORIALS

SCORE

1 $a = 33°$ Alternate angles.
$b = 82°$ Angles on a straight line.
$c = 65°$ Angles of a triangle.
$d = 33°$ Vertically opposite angles.

2 **(a)** (i) 1 line

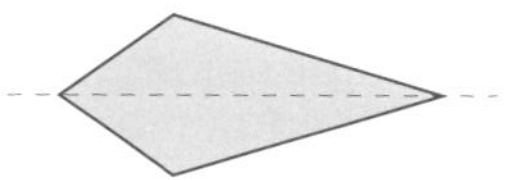

(ii) 6 lines

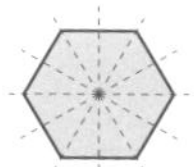

(b) (i) 4 planes

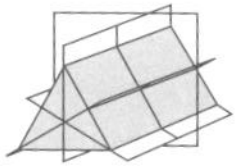

(ii) 4 planes

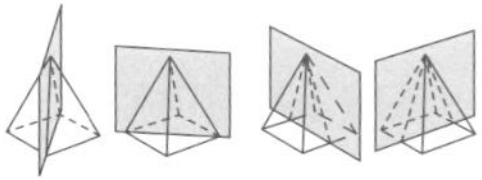

3 315°
From the diagram,
bearing = 270° + 45°
= 315°

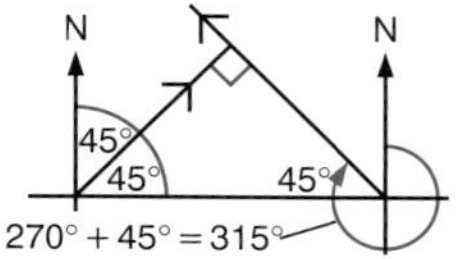

4 900°
A heptagon is a seven-sided shape, which can be split into five triangles.
Angle sum = 5 × 180° = 900°

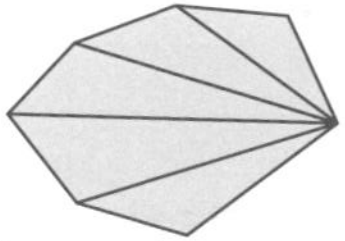

5 ∠ACO = 35° Base angles of isosceles triangle AOC.
∠ACB = 90° Angle in a semicircle is always 90°.
∠OCB = 55° 90° – 35°.

SCORE

6 **(a)** $x = 28°$
∠DBC = 62° Angles subtended by same arc DC.
$x = 180° – (62° + 90°)$ Angle sum of a triangle= 180°.
$x = 28°$

(b) ∠DPC = 118°
∠DPC is 118° because ADPC is a cyclic quadrilateral, opposite angles sum 180°.
∠DPC = 180° – ∠DAC
∠DPC = 180° – 62° = 118°

7 $A_2(4, ^-1)$,
$B_2(5, ^-1)$,
$C_2,(5\frac{1}{2}, ^-2)$,
$D_2(4\frac{1}{2}, ^-2)$

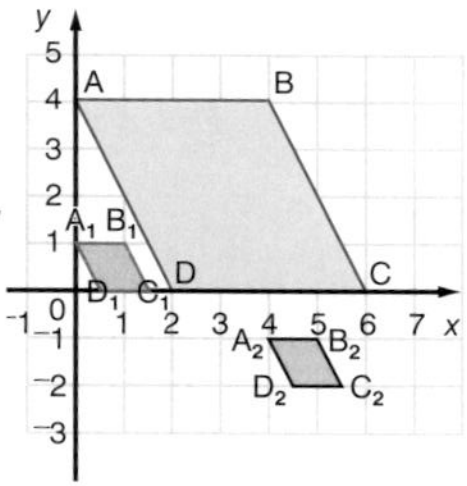

8 A reflection in the line $y = 1$
(or the line PQ)

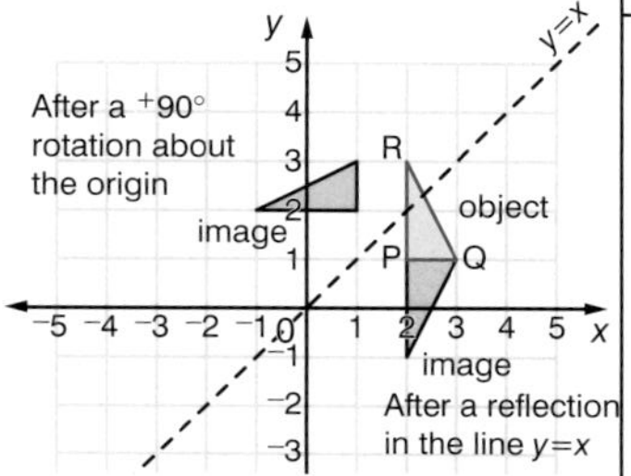

TOTAL SCORE OUT OF 20

Geometrical constructions

The following constructions can be undertaken using only a ruler and a pair of compasses.

- **The perpendicular bisector of a line (AB)**

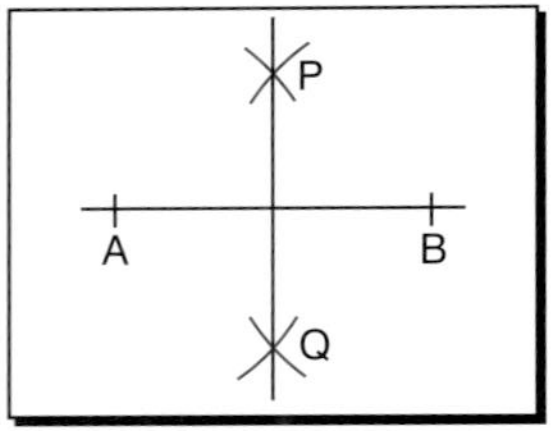

1 With the compasses set to a radius greater than half the length of AB, and centred on A, draw arcs above and below the line.
2 With the compasses still set to the same radius, and centred on B, draw arcs above and below the line, to cut the first arcs.
3 Join the points where the arcs cross (P and Q). This line is the perpendicular bisector of AB.

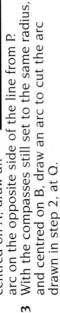

- **The perpendicular from a point (X) on a straight line**

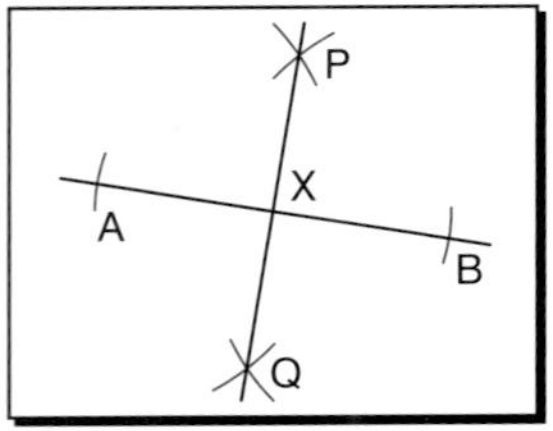

1 With the compasses set to a suitable radius greater then the distance of P from the line, and centred on X, draw arcs to cut the line at A and B.
2 Now construct the perpendicular bisector of the line segment AB.

- **The perpendicular from a point (P) to a straight line (AB)**

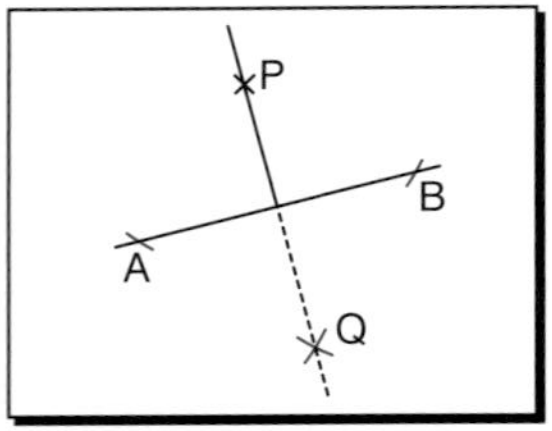

1 With compasses set to suitable radius, and centred on P, draw arcs to cut the line at A and B.
2 With the compasses set to a radius greater than half the length of AB, and centred on A, draw an arc on the opposite side of the line from P.
3 With the compasses still set to the same radius, and centred on B, draw an arc to cut the arc drawn in step 2, at Q.
4 Join PQ.

- **The angle bisector (of ∠ABC)**

1 With the compasses set to a suitable radius, for example, about 5 cm, and centred on B, draw arcs to cut BA at L and BC at M.
2 With the compasses set to the same radius, and centred on L, draw an arc between BA and BC.
3 With the compasses still set to the same radius, and centred on M, draw an arc to cut the arc between BA and BC, at Q.
4 Join BQ. This is the bisector of angle ABC.

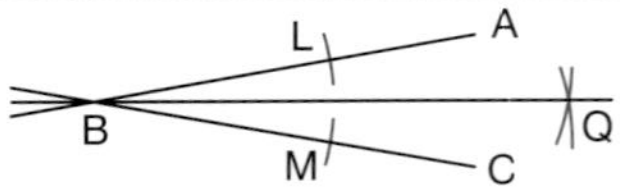

Locus of points

A **locus** is a path along which a point moves to satisfy some given rule. These are the most common examples of loci found on examination papers.

- **A point moving so that it is a fixed distance from a point O**

  The locus of a point moving so that it is a fixed distance from a point O is a circle with centre O.

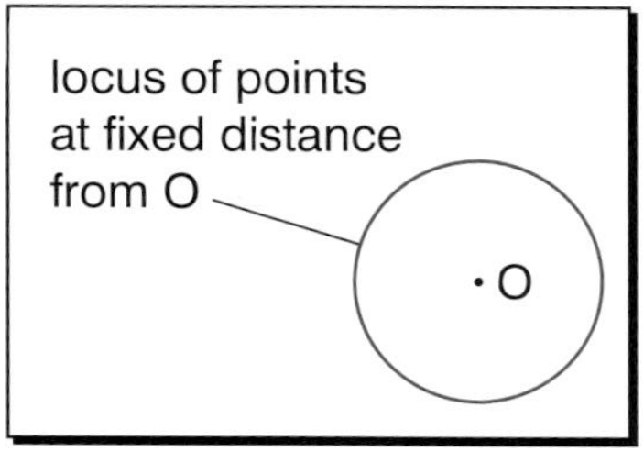

- **A point moving so that it is a fixed distance from two fixed points A and B**

  The locus of a point moving so that it is a fixed distance from two fixed points A and B is the perpendicular bisector of the line AB.

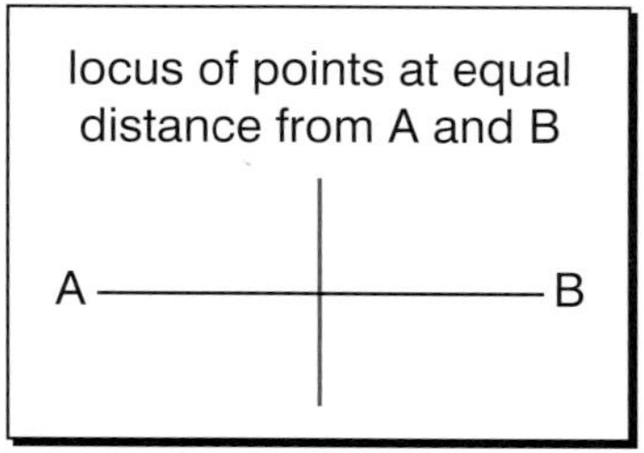

- **A point moving so that it is a fixed distance from the line PQ**

  The locus of a point moving so that it is a fixed distance from the line PQ is a pair of lines parallel to PQ, with a semicircle centred at each of the points P and Q.

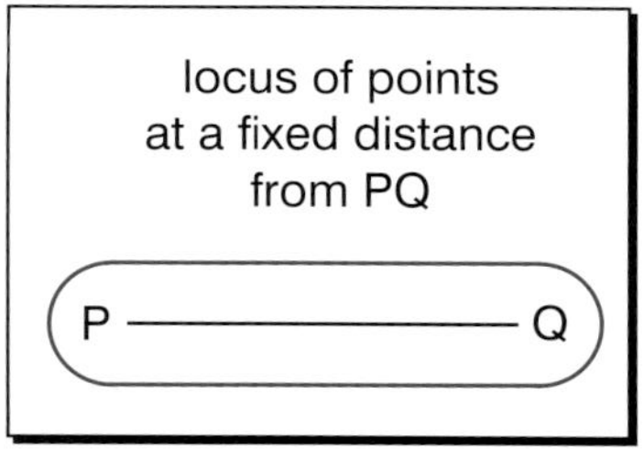

- **A point moving so that it is a fixed distance from two lines AB and CD**

  The locus of a point moving so that it is a fixed distance from two lines AB and CD is the pair of angle bisectors of the angles between the two lines (drawn from the point where the lines cross).

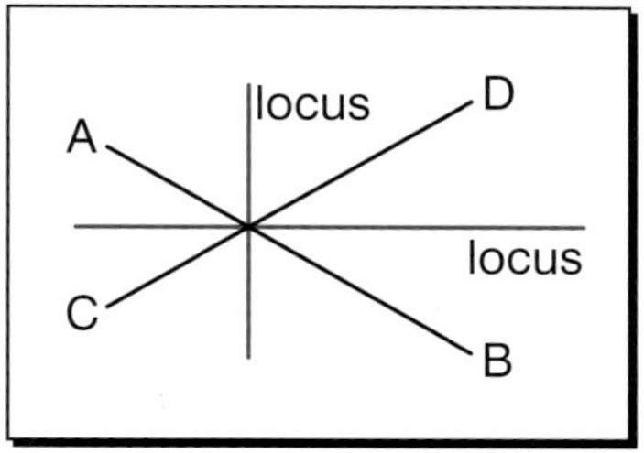

Length, area and volume

The following information will be useful for the examination paper at the Intermediate tier.

Circumference of circle

$= \pi \times$ diameter

$= 2 \times \pi \times$ radius

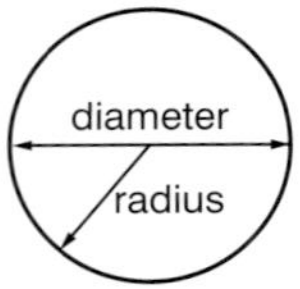

Area of circle

$= \pi \times (\text{radius})^2$

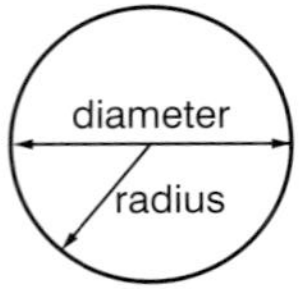

Area of parallelogram

$=$ base $\times$ height

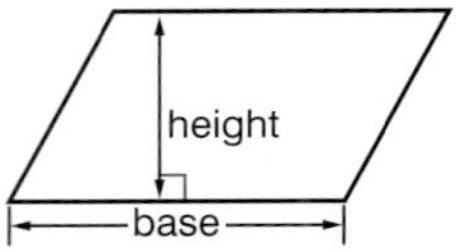

Volume of cuboid

$=$ length $\times$ width $\times$ height

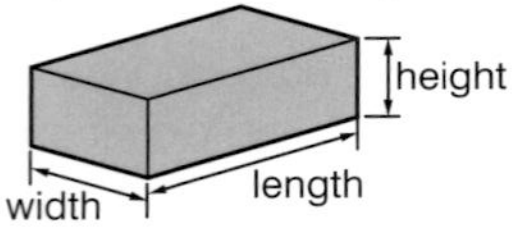

Volume of cylinder

$= \pi r^2 h$

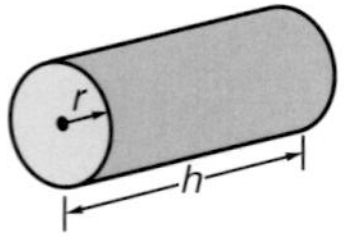

The following information will be given to you on the examination paper at the Intermediate level.

Area of trapezium

$= \frac{1}{2}(a + b)h$

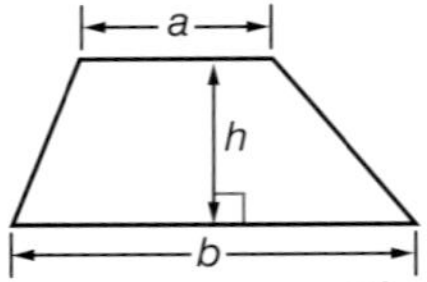

Volume of prism

$=$ area of cross-section $\times$ length

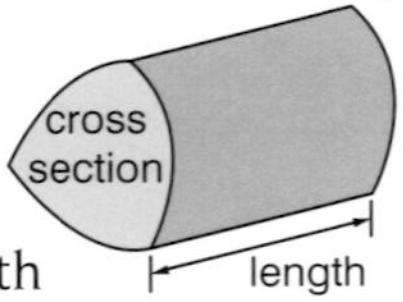

Units for length, area and volume

Lengths, areas and volumes can all be identified by looking at the formulae or units:

- area is the product of two lengths
- volume is the product of three lengths.

By ignoring constants (including π) you should be able to identify length, area and volume.

e.g. πd and $2\pi r$ are measures of length

bh, $\frac{1}{2}(a + b)h$ and πr^2 are measures of area

lbh, $4\pi r^3$, $\frac{1}{3}\pi r^2 h$ and $\frac{4}{3}\pi r^3$ are measures of volume.

Congruent triangles

Two triangles are **congruent** if one of the triangles can be fitted exactly over the other, so that all corresponding angles and corresponding sides are equal.

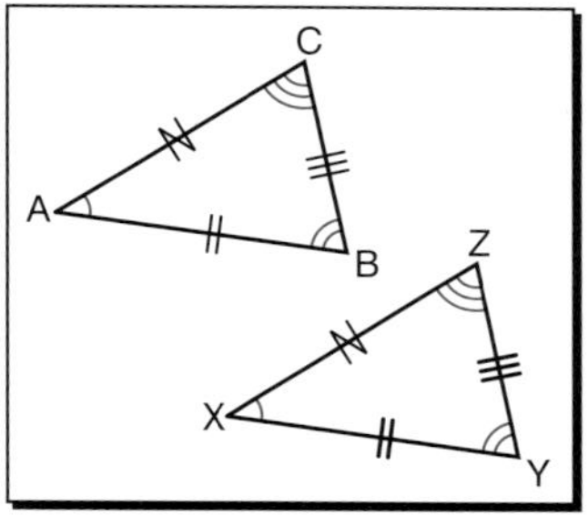

AB = XY $\angle A = \angle X$
BC = YZ $\angle B = \angle Y$
CA = ZX $\angle C = \angle Z$

If the two triangles ABC and XYZ are congruent then you can write $\Delta ABC \equiv \Delta XYZ$.

The following minimum conditions are enough to show that two triangles are congruent:

- two angles and a side of one triangle equal two angles and the corresponding side of the other (AAS)
- two sides and the included angle of one triangle equal two sides and the included angle of the other (SAS)
- the three sides of one triangle equal the three sides of the other (SSS).

If the triangles are right-angled then they are congruent if the hypotenuses are equal and two other corresponding sides are equal in the triangles (RHS).

Similar triangles

Two triangles are **similar** if one of the triangles is an enlargement of the other triangle so that all corresponding angles are equal and corresponding sides are in the same ratio.

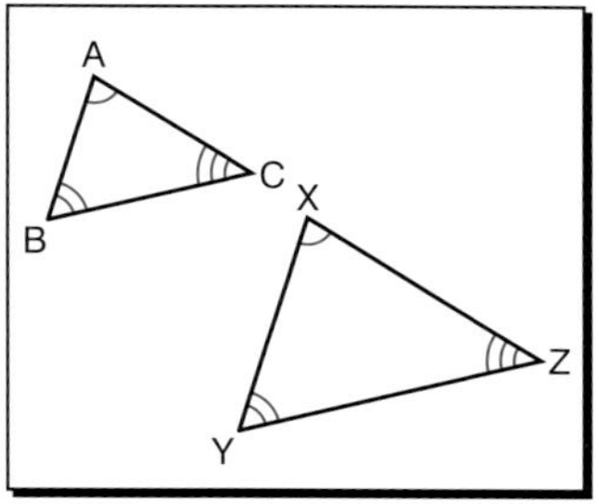

$\angle A = \angle X$
$\angle B = \angle Y$
$\angle C = \angle Z$

If two triangles are similar then the ratios of the corresponding sides are equal.

$$\frac{AB}{XY} = \frac{BC}{YZ} = \frac{CA}{ZX} \quad \text{or}$$

$$\frac{XY}{AB} = \frac{YZ}{BC} = \frac{ZX}{CA}$$

The following minimum conditions are enough to show that two triangles are similar:

- two angles of one triangle equal two angles of the other
- two pairs of sides are in the same ratio and the included angles are equal
- three pairs of sides are in the same ratio.

Pythagoras' theorem

For any right-angled triangle, the square of the length of the hypotenuse is equal to the sum of the squares of the lengths of the other two sides.

$a^2 + b^2 = c^2$

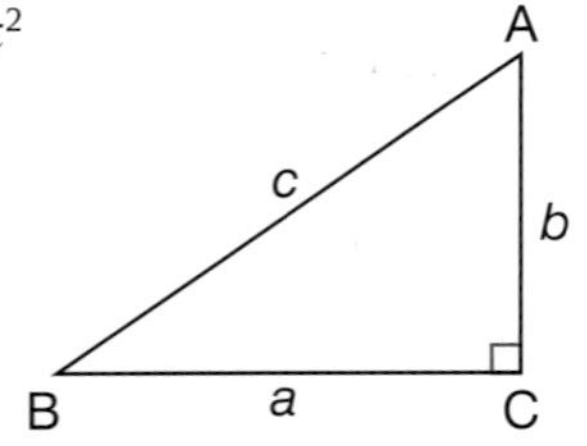

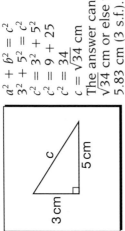

In any right-angled triangle, the side opposite the right angle is called the **hypotenuse** and this is always the longest side.

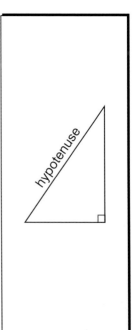

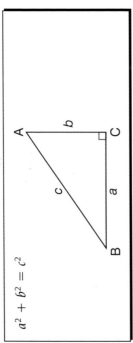

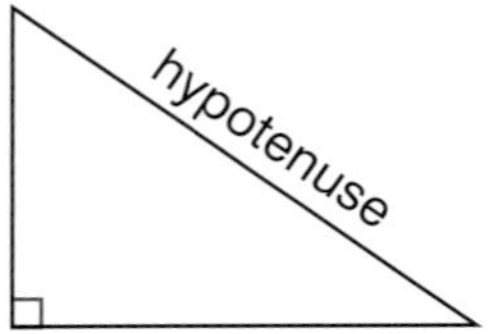

e.g. Use Pythagoras' theorem to find the missing length of the side in this right-angled triangle.

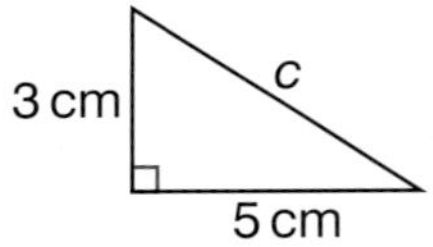

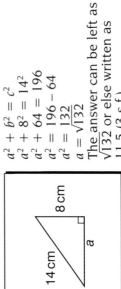

$a^2 + b^2 = c^2$
$3^2 + 5^2 = c^2$
$c^2 = 3^2 + 5^2$
$c^2 = 9 + 25$
$c^2 = 34$
$c = \sqrt{34}$ cm

The answer can be left as $\sqrt{34}$ cm or else written as 5.83 cm (3 s.f.).

e.g. Use Pythagoras' theorem to find the missing length of the side in this right-angled triangle.

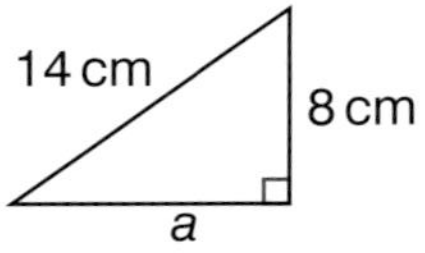

$a^2 + b^2 = c^2$
$a^2 + 8^2 = 14^2$
$a^2 + 64 = 196$
$a^2 = 196 - 64$
$a^2 = 132$
$a = \sqrt{132}$

The answer can be left as $\sqrt{132}$ or else written as 11.5 (3 s.f.).

On a non-calculator paper you should usually leave your answer in surd form. On a calculator paper you should normally round your answers to an appropriate degree of accuracy (e.g. 3 s.f.).

Trigonometry

The sides of a right-angled triangle are given special names, as shown in this diagram.

For angle A:

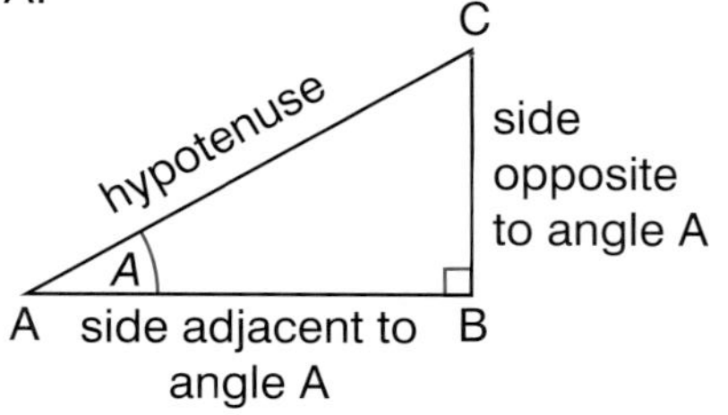

For angle C:

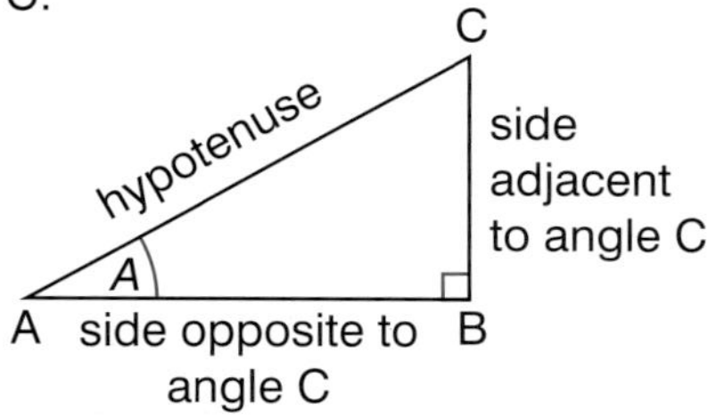

Sine of an angle

The sine of an angle (usually abbreviated as sin)

$$= \frac{\text{length of opposite side}}{\text{length of hypotenuse}}$$

So $\quad \sin A = \dfrac{\text{length of side opposite to A}}{\text{length of hypotenuse}} = \dfrac{\text{BC}}{\text{AC}}$

and $\quad \sin C = \dfrac{\text{length of side opposite to C}}{\text{length of hypotenuse}} = \dfrac{\text{AB}}{\text{AC}}$

Cosine of an angle

The cosine of an angle (usually abbreviated as cos)

$$= \frac{\text{length of adjacent side}}{\text{length of hypotenuse}}$$

So $\quad \cos A = \dfrac{\text{length of side adjacent to A}}{\text{length of hypotenuse}} = \dfrac{\text{AB}}{\text{AC}}$

and $\quad \cos C = \dfrac{\text{length of side adjacent to C}}{\text{length of hypotenuse}} = \dfrac{\text{BC}}{\text{AC}}$

Tangent of an angle

The tangent of an angle (usually abbreviated as tan)

$$= \frac{\text{length of opposite side}}{\text{length of adjacent side}}$$

So $\tan A = \dfrac{\text{length of side opposite to A}}{\text{length of side adjacent to A}} = \dfrac{BC}{AB}$

So $\tan C = \dfrac{\text{length of side opposite to C}}{\text{length of side adjacent to C}} = \dfrac{AB}{BC}$

Using sine, cosine and tangent

Use the sine, cosine and tangent ratios to find missing lengths.

Finding lengths

e.g. Find the length c in this right-angled triangle.

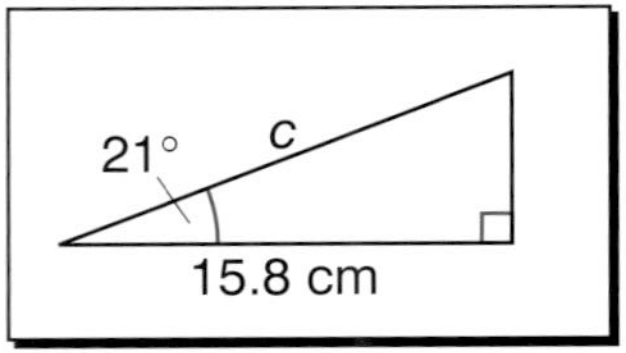

$$\cos B = \frac{\text{length of adjacent side}}{\text{length of hypotenuse}}$$

$$\cos 21° = \frac{15.8}{c}$$

$$c \times \cos 21° = 15.8$$

$$c = \frac{15.8}{\cos 21°}$$

$$c = \frac{15.8}{0.933\,580\,4}$$

$$c = 16.9\text{ cm (3 s.f.)}$$

e.g. Find the area of this rectangle.

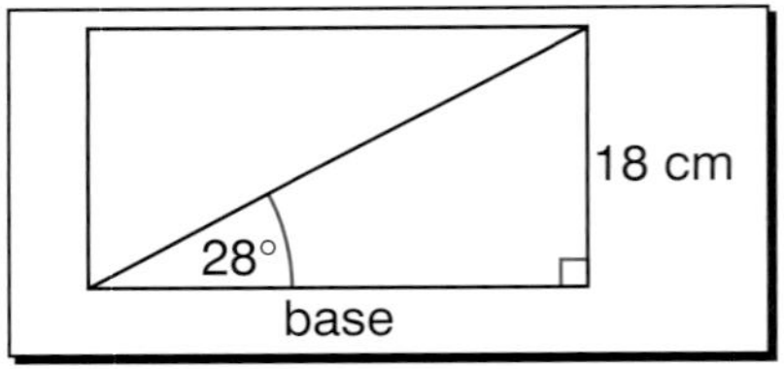

Area of rectangle = base × perpendicular height

To find base use $\tan A = \dfrac{\text{length of opposite side}}{\text{length of adjacent side}}$

$$\tan 28° = \frac{18}{\text{base}}$$

$$\text{base} = \frac{18}{\tan 28°}$$

$$\text{base} = \frac{18}{0.531\,709\,4}$$

$$\text{base} = 33.853\,076\text{ cm}$$

Area of rectangle = base × perpendicular height

Area of rectangle = $33.853\,076 \times 18$
= $609.355\,37$
= 609 cm^2 (3 s.f.)

rounding the final answer to an appropriate degree of accuracy and remembering to include the units.

Finding angles

You can find the angles of a right-angled triangle by using the $\sin^{-1}$ (or arcsin), $\cos^{-1}$ (or arccos) and $\tan^{-1}$ (or arctan) buttons on your calculator.

First you need to find the sine, cosine or tangent of the angle, then you can use the inverse button to find the angle.

e.g. Find the angles marked α and β in the following right-angled triangles.

(a)

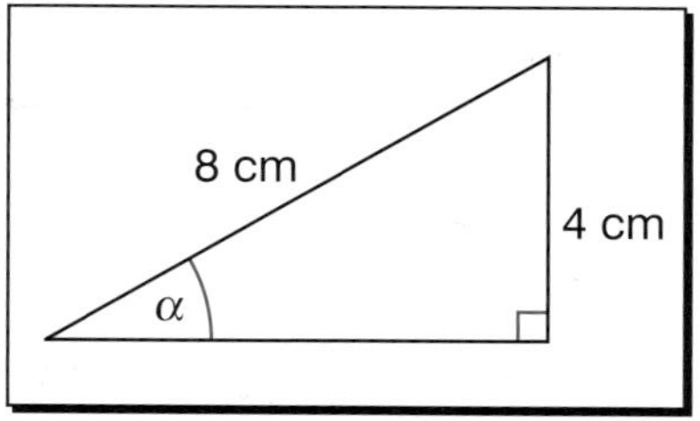

$$\sin\alpha = \frac{\text{length of opposite side}}{\text{length of hypotenuse}}$$
$$\sin\alpha = \frac{4}{8}$$
$$\sin\alpha = 0.5$$
$$\alpha = \sin^{-1} 0.5$$
$$\alpha = 30°$$

(b)

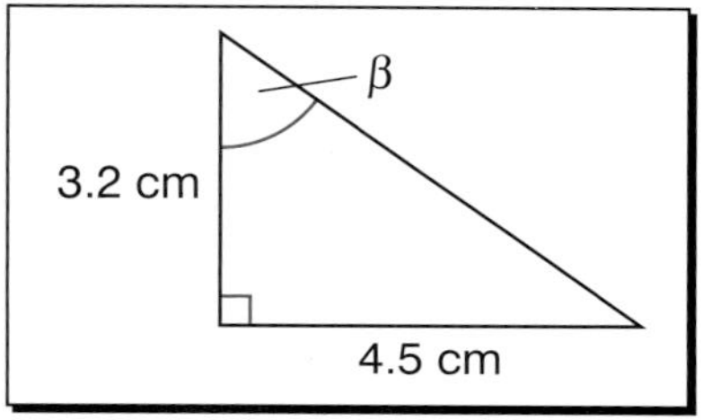

$$\tan\beta = \frac{\text{length of opposite side}}{\text{length of adjacent side}}$$
$$\tan\beta = \frac{4.5}{3.2}$$
$$\tan\beta = 1.406\,25$$
$$\beta = \tan^{-1} 1.406\,25$$
$$\beta = 54.6° \text{ (3 s.f.)}$$

Elevation and depression

The angle of elevation is the angle up from the horizontal.

The angle of depression is the angle down from the horizontal.

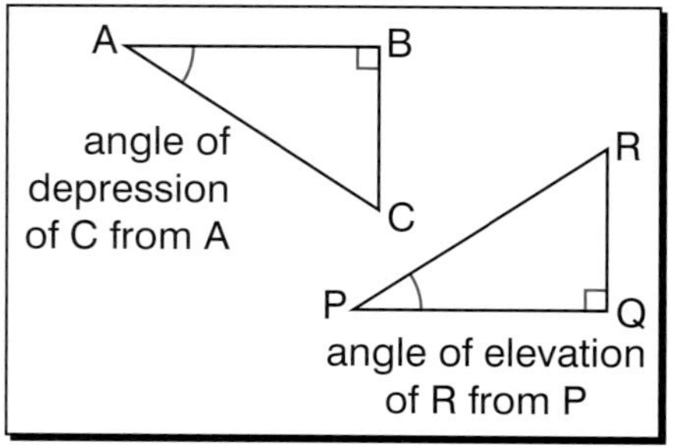

Shape, space & measures 12–19

1 Calculate the perimeter and area of each of these shapes.

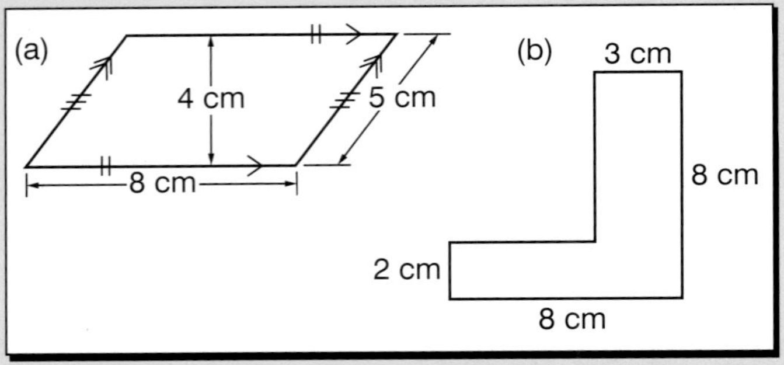

(1 *mark for each correct part*)

2 Calculate the circumference and area of a circle with diameter 8 cm. Leave your answer in surd form.

(2 *marks*)

3 A ship sails 15 km on a bearing of 045°, then 20 km on a bearing of 135°. How far is the ship away from its starting point?

(2 *marks*)

4 Draw the line AC so that AC = 6 cm.
Construct the perpendicular bisector of AC.
Construct the points B and D so that they lie on the perpendicular bisector at a distance of 4 cm from the line AC.
Join AB, BC, CD and DA.
What is the special name given to this quadrilateral?

(2 *marks for construction and* 1 *mark for name*)

5 From the top of a vertical cliff 525 metres high, the angle of depression of a boat at sea is 42°. What is the distance of the boat from the foot of the cliff?

(2 *marks*)

6 A rectangle measures 12 cm by 8 cm. A similar rectangle has sides of 6 cm and x cm. What are the possible values of x?

(1 *mark for each value of x*)

7 A cylinder has a radius of 4.1 cm and a volume of 380 ml.
Calculate the height of the cylinder, giving your answer correct to an appropriate degree of accuracy.

(2 *marks*)

8 Identify the following as length, area or volume.

(a) $2\pi r(r + h)$ **(b)** $\frac{\theta}{360} \times \pi d$ **(c)** $\frac{4}{3}\pi r^3$

(1 *mark for each part*)

Total marks = 20

ANSWERS & TUTORIALS

1 **(a)** Perimeter = 26 cm
Perimeter = 8 + 5 + 8 + 5 = 26 cm.
Area = 32 cm^2 Area = $8 \times 4 = 32$ cm^2.
(b) Perimeter = 32 cm Perimeter = 8 + 8 + 3 + 6 + 5 + 2 = 32 cm.
Area = 34 m^2 Area = $2 \times 5 + 3 \times 8$ = 34 m^2 (two rectangles).

2 Perimeter = $\pi \times d = \pi \times 8$
= 8π cm
Area = πr^2
= $\pi \times 4^2$
= 16π cm^2

3 Distance = 25 km
Using Pythagoras' theorem:
$a^2 + b^2 = c^2$
$15^2 + 20^2 = c^2$
$c = 25$ km

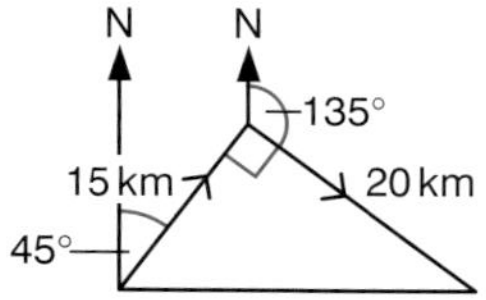

4 Rhombus

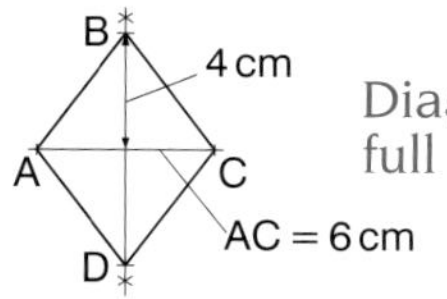

Diagram not full size

5 583 m (3 s.f.)
$\tan 42° = \frac{525}{x}$
$x = \frac{525}{\tan 42°}$ $x = 583.071\,57$ m

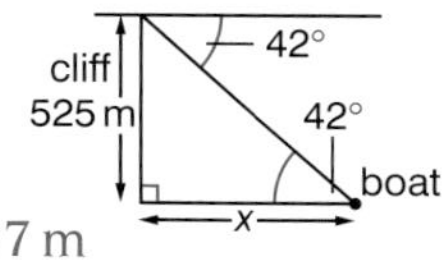

6 $x = 4$ cm or $x = 9$ cm

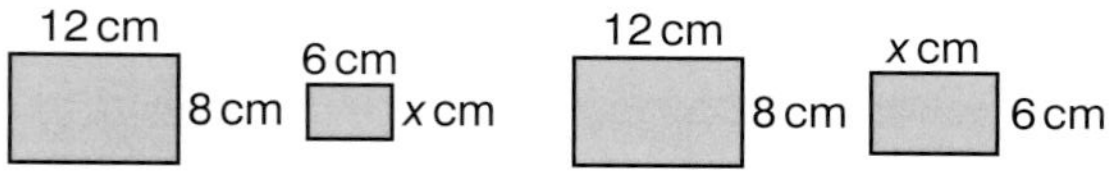

$\frac{12}{8} = \frac{6}{x}$ $\frac{12}{8} = \frac{x}{6}$

$x = \frac{8}{12} \times 6 = 4$ cm $x = \frac{12}{8} \times 6 = 9$ cm

7 7.2 cm to an appropriate degree of accuracy
$V = \pi r^2 h$
$380 = \pi \times 4.1^2 \times h$
380 ml is equivalent to 380 cm^3.
$h = \frac{380}{\pi \times 4.1^2} = 7.195\,583\,4$ cm
= 7.2 cm to an appropriate degree of accuracy.

8 **(a)** Area
In $2\pi r(r + h)$ 2π is a constant, $r + h$ = length and $r(r + h)$ gives length × length = area.
(b) Length
$\frac{\theta}{360} \times \pi$ is a constant and d gives the unit of length.
(c) Volume
$\frac{4}{3}\pi$ is a constant and r^3 gives units of length × length × length = volume.

SCORE

SCORE

TOTAL SCORE OUT OF 20

Shape, space & measures 12–19

1 Calculate the volume of each of these solids.

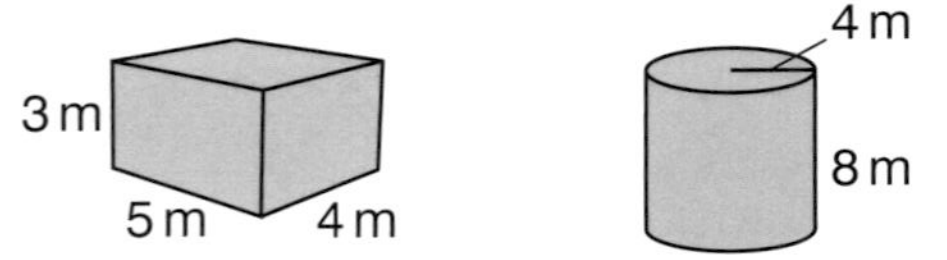

(*2 marks for each part*)

2 Calculate the surface area of a cube of side 3.5 cm.

(*2 marks*)

3 Draw the locus of points less than 2 cm from a square of side 4 cm.

(*2 marks*)

4 ABC is a right-angled triangle with $\angle BCA = 90°$. If AC = 4 cm and BC = 6 cm what is the perimeter of the triangle?

(*2 marks*)

5 A pylon of height 50 m casts a shadow which is 67 m long. Calculate the angle of elevation of the sun.

(*2 marks*)

6 The following diagram shows the cross-section of a swimming pool. Calculate the greatest depth of the pool.

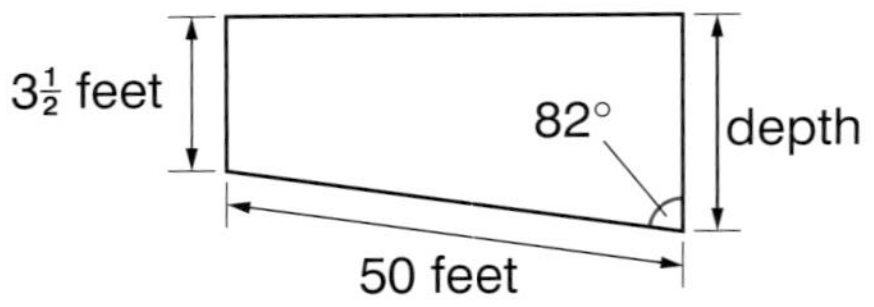

(*3 marks*)

7 Prove that $\Delta PQT \equiv \Delta SRT$ and hence find the length PQ.

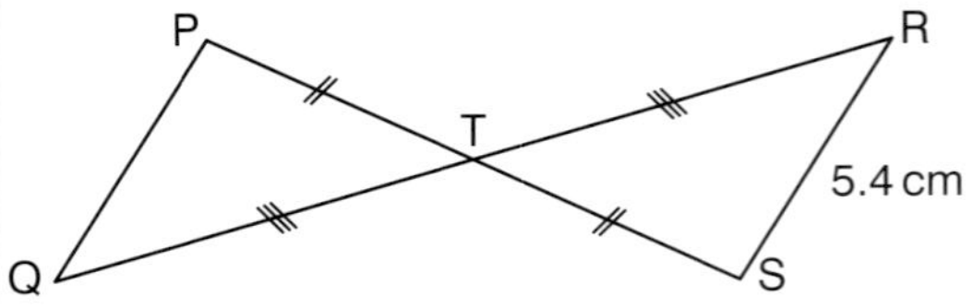

(*2 marks*)

8 A metal block has a triangular cross-section with measurements as shown on the diagram.
Find the area of the cross-section and the volume of the block.

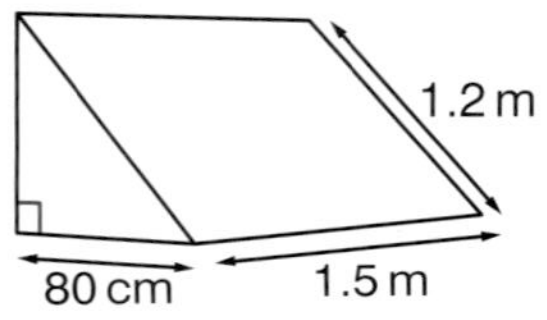

(*2 marks for the area and 1 mark for the volume*)

***Total marks* = 20**

ANSWERS & TUTORIALS

1 (a) 60 m^3

Volume $= lbh = 5 \times 4 \times 3 = 60\text{ m}^3$

(b) 402 m^3 (3 s.f.)

Volume $= \pi r^2 h = \pi \times 4^2 \times 8$

$= 402.123\,86\text{ m}^3$

2 73.5 cm^2

Surface area $= 6 \times 3.5^2 = 73.5\text{ cm}^2$

3

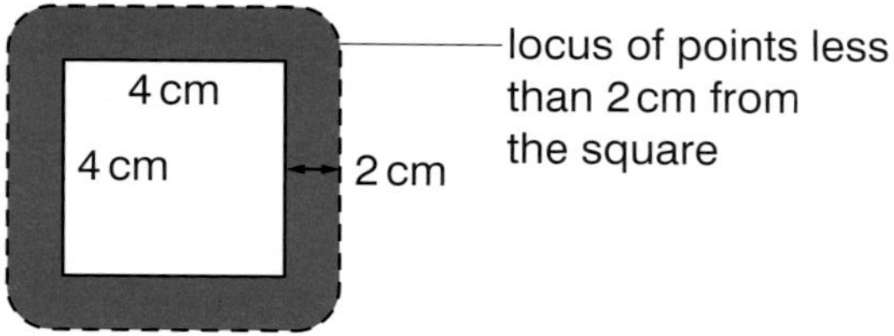

4 $10 + \sqrt{52}$ cm

Using Pythagoras':

$a^2 + b^2 = c^2$

$4^2 + 6^2 = c^2$

$c^2 = 52$

$c = \sqrt{52}$

Perimeter $= 4 + 6 + \sqrt{52}$

$= 10 + \sqrt{52}$ cm

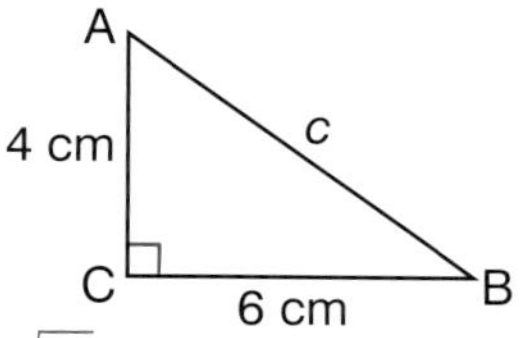

5 $36.7°$ (3 s.f.)

$\tan\theta = \frac{50}{67}$

$\theta = 36.732\,827°$

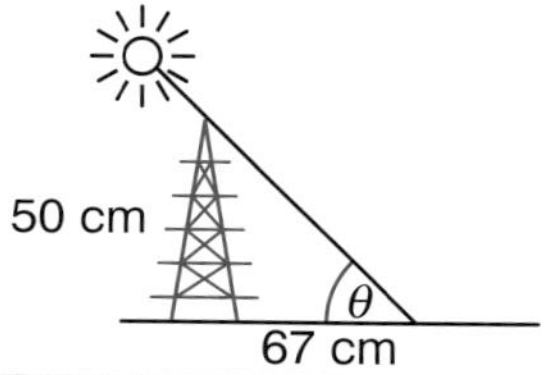

6 10.5 feet (3 s.f.)

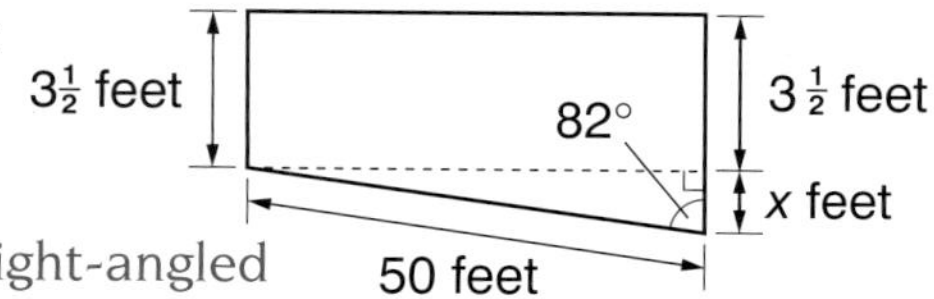

For the right-angled triangle, $\cos 82° = \frac{x}{50}$ $\quad x = 6.958\,655$

Depth $= 6.958\,655 + 3.5 = 10.458\,655$ feet

7 PQ = 5.4 cm

$\angle PTQ = \angle RTS$	Vertically opposite angles.
TQ = TR	Given
TP = TS	Given
so $\Delta PQT \equiv \Delta SRT$	SAS

As $\Delta PQT \equiv \Delta SRT$ then PQ = RS = 5.4 cm

8 Area $= 3580\text{ cm}^2$ (3 s.f.)

or 0.358 m^2 (3 s.f.)

Volume $= 537\,000\text{ cm}^3$ (3 s.f.)

or 0.537 m^3 (3 s.f.)

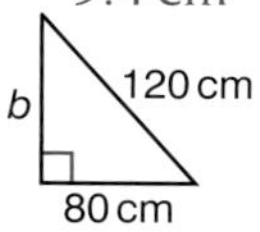

$80^2 + b^2 = 120^2$ $\quad$ Pythagoras' theorem.

$6400 + b^2 = 14\,400$ $\quad$ Converting 1.2 m to cm.

$b^2 = 8000$ so $b = 89.442\,719$

Area of cross-section

$= \frac{1}{2} \times$ base $\times$ perpendicular height

$= \frac{1}{2} \times 80 \times 89.442719 = 3577.7088\text{ cm}^2$

Volume of block

$=$ area of cross-section $\times$ length

$= 3577.7088 \times 150$

$= 536\,656.31\text{ cm}^3$

TOTAL SCORE OUT OF 20

Sine, cosine and tangent

The sine, cosine and tangent for any angle can be found from a calculator or by using the properties of sine, cosine and tangent curves as shown in these graphs.

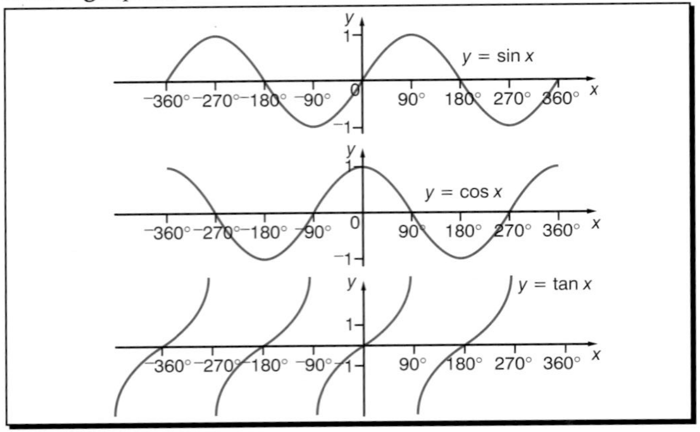

A calculator gives the sine, cosine and tangent of any angle although the reverse process only gives answers in a specified range.

e.g. Find the values of θ in the range $^-360° \leqslant \theta \leqslant {}^+360°$ such that $\tan\theta = 1$.

From a calculator, $\tan\theta = 1$ gives $\theta = 45°$.

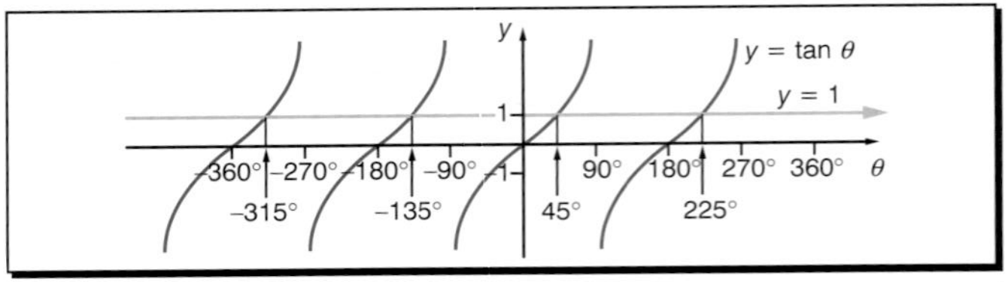

From the graph, all the solutions in the given range $^-360° \leqslant \theta \leqslant +360°$ are $^-315°$, $^-135°$, $45°$ and $225°$.

NB Check these on your calculator by evaluating $\tan {}^-315°$, $\tan {}^-135°$ and $\tan 225°$.

Cosine rule

The cosine rule is used to solve non-right angled triangles involving three sides and one angle. The cosine rule states:

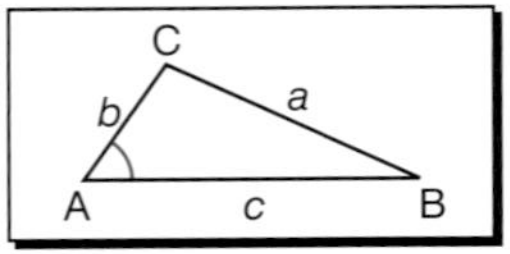

$$a^2 = b^2 + c^2 - 2bc\cos A$$

$$\text{or } \cos A = \frac{b^2 + c^2 - a^2}{2bc}$$

Sine rule

The sine rule is used to solve non-right angled triangles involving two sides and two angles. The sine rule states:

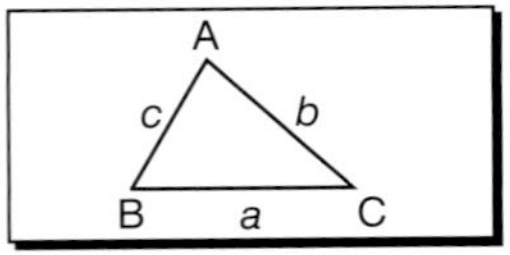

$$\frac{a}{\sin A} = \frac{b}{\sin B} = \frac{c}{\sin C}$$

Using the sine and cosine rules

e.g. Use the sine rule to evaluate angle A in this triangle.

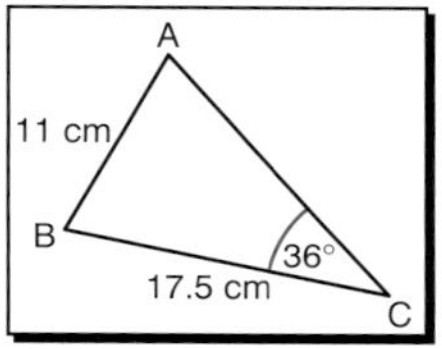

$\frac{\sin A}{a} = \frac{\sin B}{b}$ Reciprocating both sides.

$\frac{\sin A}{17.5} = \frac{\sin 36°}{11}$

$\sin A = 0.935\,112\,9$

$A = 69.246\,393°$ Using the inverse button ($\sin^{-1}$ or arcsin) on the calculator.

Unfortunately the value of $A = 69.246\,393°$ is not unique since there is another possible value for A which satisfies $\sin A = 0.935\,112\,9$.

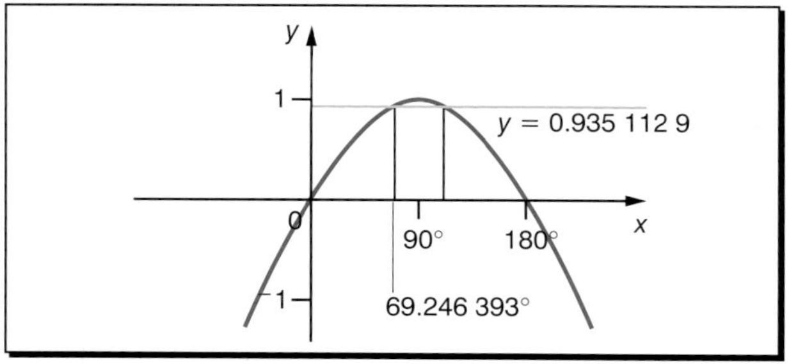

From the graph, another possible value of A is $110.753\,61°$.

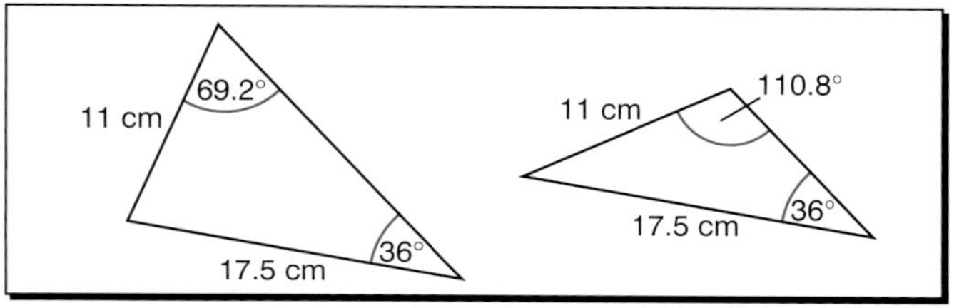

So $A = 69.2°$ or $110.8°$ (1 d.p.)

NB This problem does not arise when using the cosine rule; it can be avoided when using the sine rule by finding the smallest angles first (if possible).

Area of a triangle

The area of triangle ABC $= \frac{1}{2}ab\sin C$

Area $= \frac{1}{2} \times$ base $\times$ height

$= \frac{1}{2}ah$

But $h = b\sin C$

So area $= \frac{1}{2}ab\sin C$

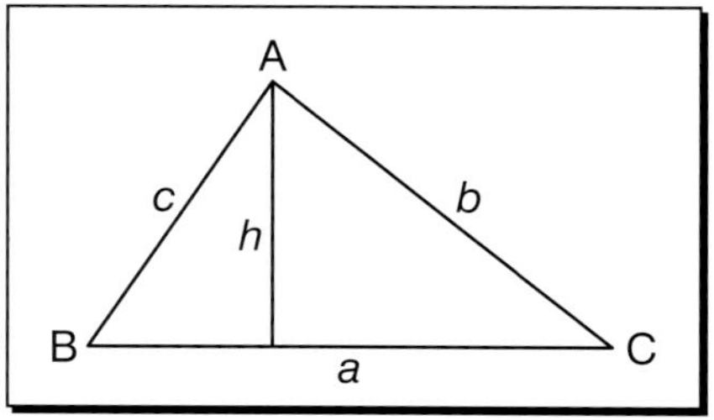

NB The angle C is the **included** angle between the sides a and b.

Further length, area and volume

The following information will be needed for the examination paper at the Higher tier.

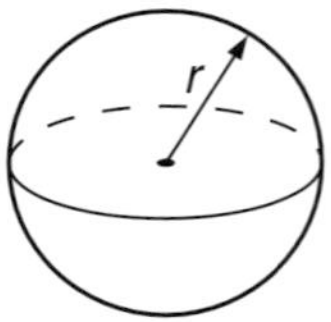

Volume of sphere $= \frac{4}{3}\pi r^3$

Surface area of sphere $= 4\pi r^2$

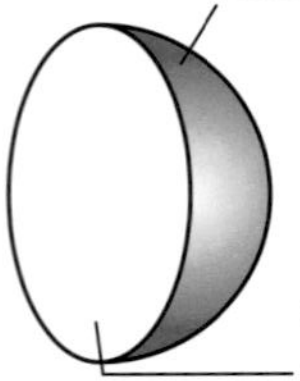

Volume of hemisphere $= \frac{1}{2} \times \frac{4}{3}\pi r^3$

$= \frac{2}{3}\pi r^3$

Surface area of hemisphere $= \frac{1}{2} \times 4\pi r^2 + \pi r^2$

$= 3\pi r^2$

Volume of cone $= \frac{1}{3}\pi r^2 h$

Surface area of cone $= \pi r l$

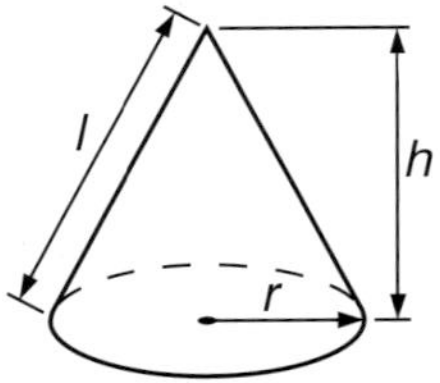

Using Pythagoras' theorem:

$l^2 = r^2 + h^2$

Scale factors

Two solids are **similar** if the ratios of their corresponding linear dimensions are equal.

In general:

- the corresponding areas of similar solids are proportional to the squares of their linear dimensions
- the corresponding volumes of similar solids are proportional to the cubes of their linear dimensions.

This means that if the ratio of the lengths is $x : y$

then the ratio of the areas is $x^2 : y^2$

and the ratio of the volumes is $x^3 : y^3$

Arc, sector and segment

The following definitions will be useful for this work.

An **arc** is a part of the circumference of a circle.

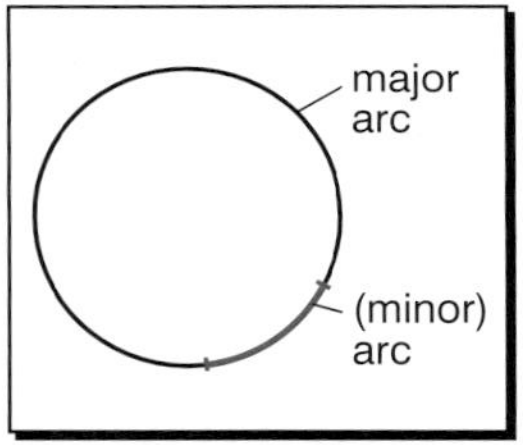

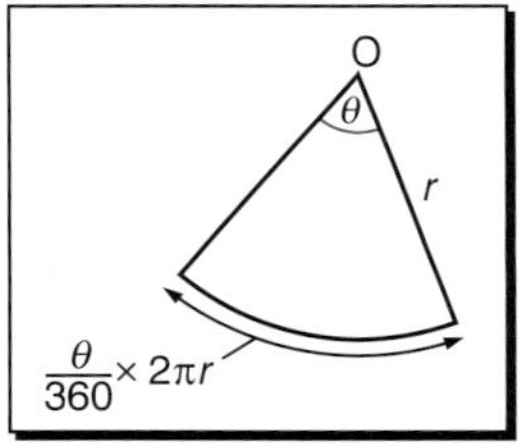

$$\text{Arc length} = \frac{\text{angle subtended at centre}}{360} \times 2\pi r$$

$$= \frac{\theta}{360} \times 2\pi r$$

A **sector** is the area enclosed between an arc and two radii.

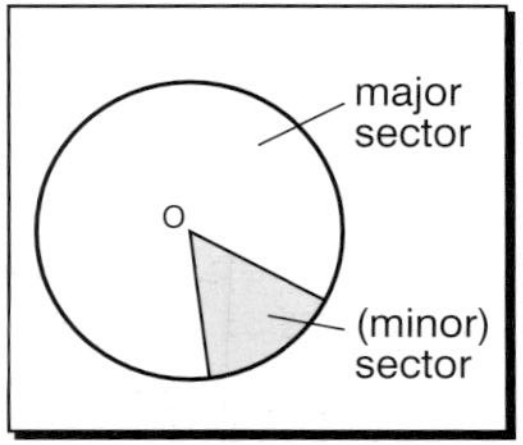

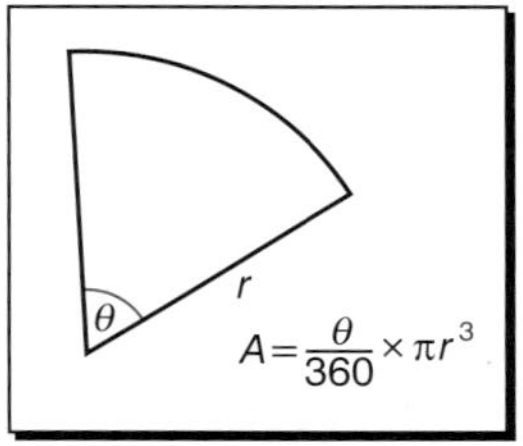

$$\text{Sector area} = \frac{\text{angle subtended at centre}}{360} \times \pi r^2$$

$$= \frac{\theta}{360} \times \pi r^2$$

A **segment** is the area enclosed between an arc and a chord.

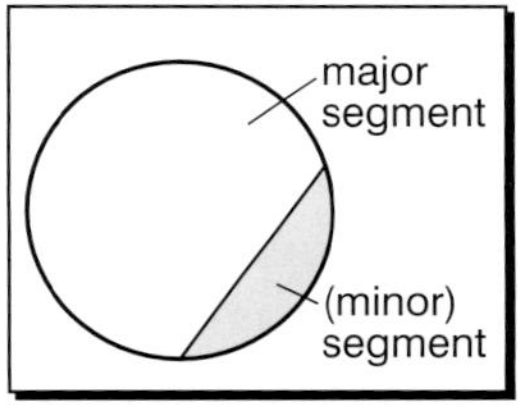

e.g. Find the area of the shaded segment in this circle of radius 6 cm.

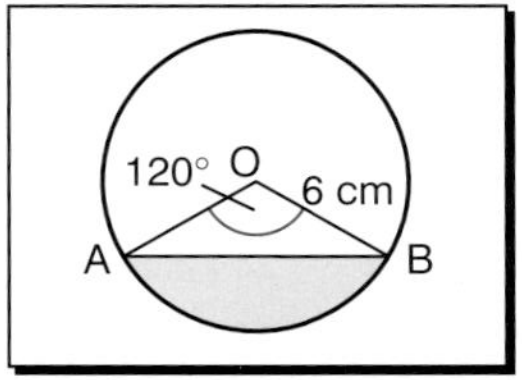

Area of segment
= area of sector AOB – area of triangle AOB

$$\text{Area of sector AOB} = \frac{120}{360} \times \pi \times 6^2 = 37.699\,112\ \text{cm}^2$$

$$\text{Area of triangle AOB} = \tfrac{1}{2}ab\sin\theta = \tfrac{1}{2} \times 6 \times 6 \times \sin 120^\circ$$
$$= 15.588\,457\ \text{cm}^2$$

Area of segment = area of sector – area of triangle

$$= 37.699\,112 - 15.588\,457$$
$$= 22.110\,655\ \text{cm}^2$$
$$= 22.1\ \text{cm}^2 \text{ (to 3 s.f.)}$$

Chords in a circle

A **chord** is a straight line joining two points on the circumference of a circle.

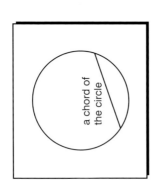

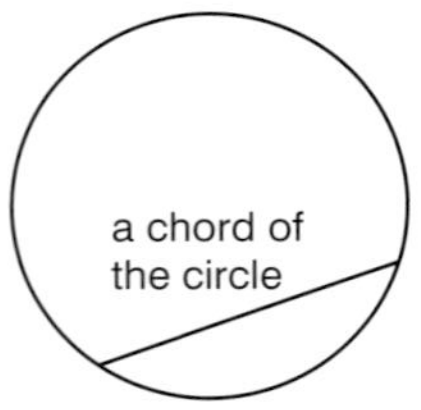

A **diameter** is a chord which passes through the centre of the circle.

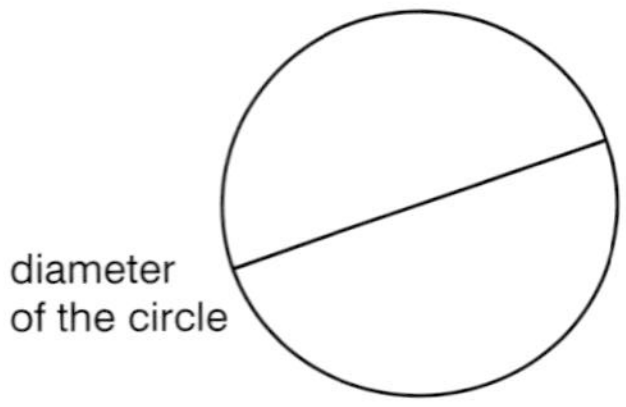

The following chord properties need to be known.

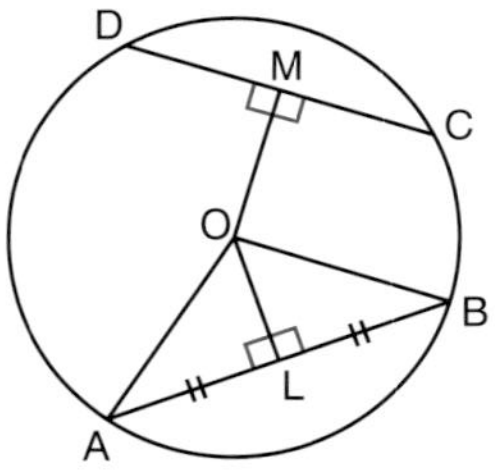

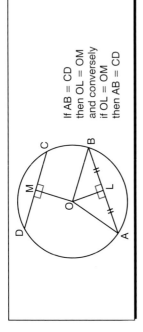

If AB = CD
then OL = OM
and conversely
if OL = OM
then AB = CD

- A perpendicular from the centre of a circle to a chord bisects the chord.

  NB Conversely, a perpendicular bisector of a chord passes through the centre of the circle.

- Chords which are equal in length are equidistant from the centre of the circle.

  NB Conversely, chords which are equidistant from the centre of a circle are equal in length.

Alternate segment theorem

The alternate angle theorem states that the angle between a tangent and a chord equals the angle subtended by the chord in the alternate segment.

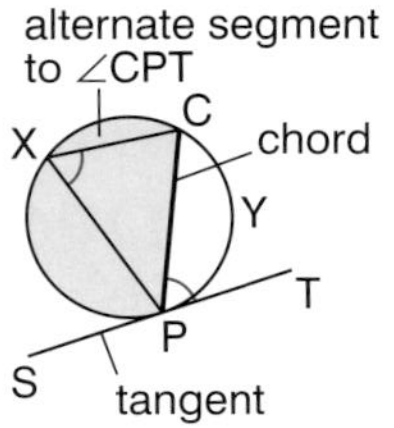

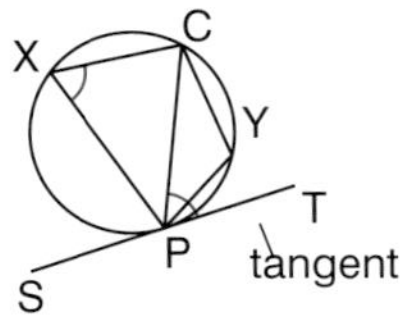

∠CPT = ∠CXP
(alternate segment theorem)

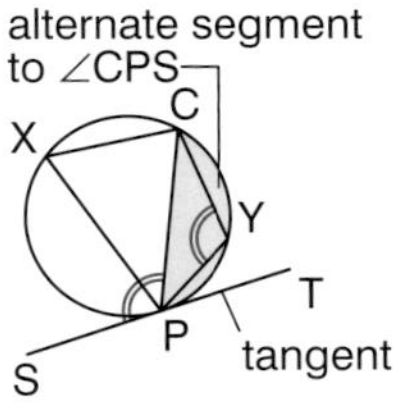

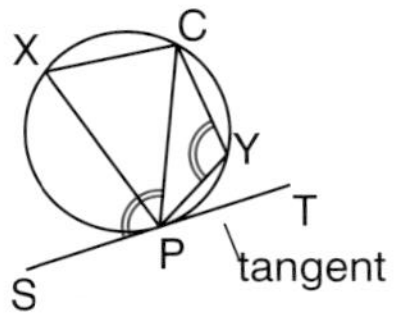

∠CPS = ∠CYP
(alternate segment theorem)

e.g. Find the size of ∠XPC, given that ST is a tangent to the circle at P.

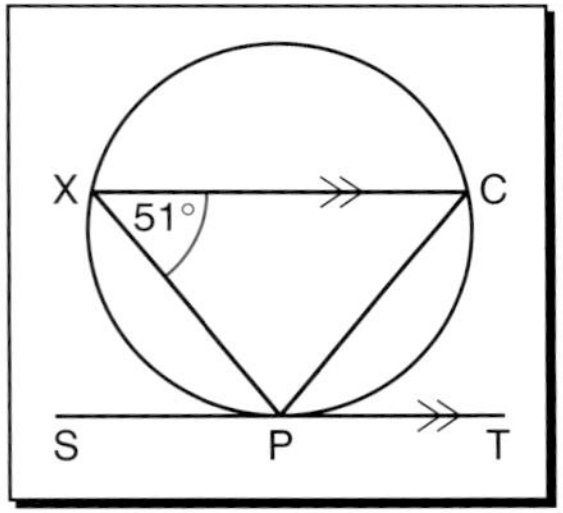

Using the alternate segment theorem:
∠CPT = ∠CXP = 51°
By the alternate segment theorem.
∠XCP = ∠CPT = 51°
Alternate angles between parallel lines XC and PT.
∠XPC = 180° – (51° + 51°)
Angles of triangle XCP add up to 180°.
∠XCP = 78°

Vectors and vector properties

A vector is a quantity which has magnitude (length) and direction (usually indicated by an arrow).

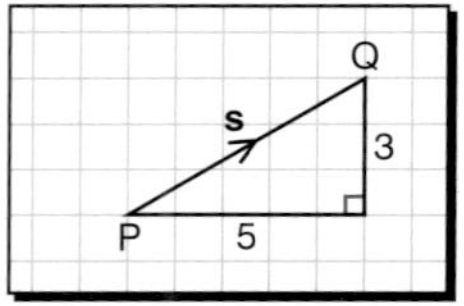

Vectors are used to define translations and also can be used to prove geometrical theorems.

The above vector can be represented as **PQ**, or $\overrightarrow{PQ}$ or **s** (you can write s̲) or as a column vector $\begin{pmatrix}5\\3\end{pmatrix}$.

Two vectors are equal if they have the same magnitude and direction, i.e. they are the same length and they are parallel.

Addition and subtraction of vectors

Vectors can be added or subtracted by placing them end-to-end so that the arrows point in the same direction.

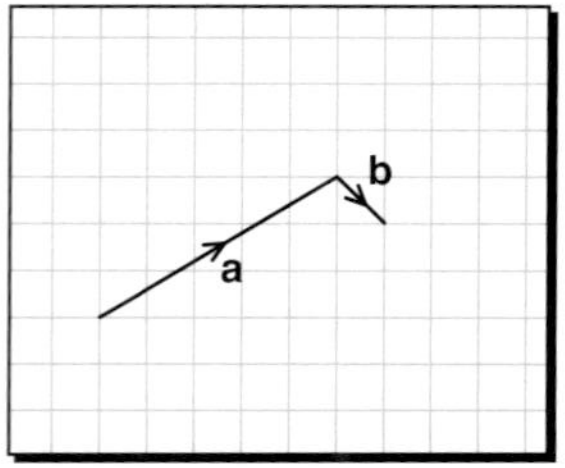

An easier way is to write them as column vectors.

If $\mathbf{a} = \begin{pmatrix}5\\3\end{pmatrix}$ and $\mathbf{b} = \begin{pmatrix}1\\-1\end{pmatrix}$:

$$\mathbf{a} + \mathbf{b} = \begin{pmatrix}5\\3\end{pmatrix} + \begin{pmatrix}1\\-1\end{pmatrix} = \begin{pmatrix}6\\2\end{pmatrix} \text{ and}$$

$$\mathbf{a} - \mathbf{b} = \begin{pmatrix}5\\3\end{pmatrix} - \begin{pmatrix}1\\-1\end{pmatrix} = \begin{pmatrix}4\\4\end{pmatrix}$$

as shown in the diagrams.

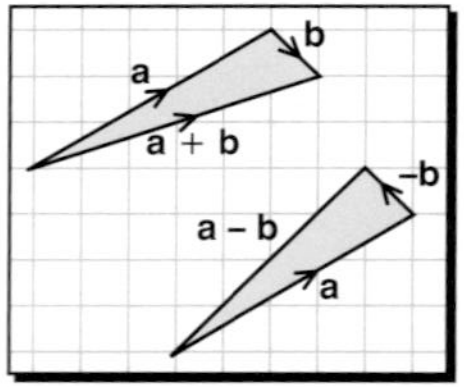

Magnitude of a vector

The magnitude (length) of a vector can be found by using Pythagoras' theorem.

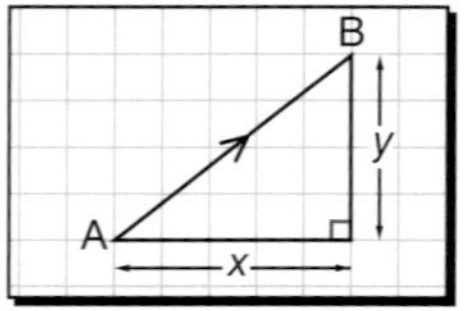

The length of the vector **AB** is $\sqrt{x^2 + y^2}$ and you can write $|\mathbf{AB}| = \sqrt{x^2 + y^2}$ where the two vertical lines mean the magnitude or length.

Multiplication of a vector

Vectors cannot be multiplied by other vectors but they can be multiplied by a constant (sometimes called **scalar multiplication**).

e.g. Given that $\mathbf{p} = \begin{pmatrix}1\\-8\end{pmatrix}$ and $\mathbf{q} = \begin{pmatrix}-2\\-5\end{pmatrix}$:

$$3\mathbf{p} = 3 \times \begin{pmatrix}1\\-8\end{pmatrix} = \begin{pmatrix}3\\-24\end{pmatrix} \text{ and } 4\mathbf{q} = 4 \times \begin{pmatrix}-2\\-5\end{pmatrix} = \begin{pmatrix}-8\\-20\end{pmatrix}$$

$$\text{so } 3\mathbf{p} - 4\mathbf{q} = \begin{pmatrix}3\\-24\end{pmatrix} - \begin{pmatrix}-8\\-20\end{pmatrix} = \begin{pmatrix}11\\-4\end{pmatrix}$$

Further enlargement

The subject of enlargement is introduced on Shape, space and measures card 9. The topic is extended to include negative scale factors at the Higher tier.

Negative scale factors

For an enlargement with a negative scale factor, the enlargement is situated on the opposite side of the centre of enlargement.

e.g. The quadrilateral PQRS with vertices P($^-1$, $^-2$), Q($^-3$, $^-2$), R($^-3$, 1) and S($^-1$, 3) is enlarged with a scale factor of $^-2$ about the origin. Draw PQRS and hence P'Q'R'S'.

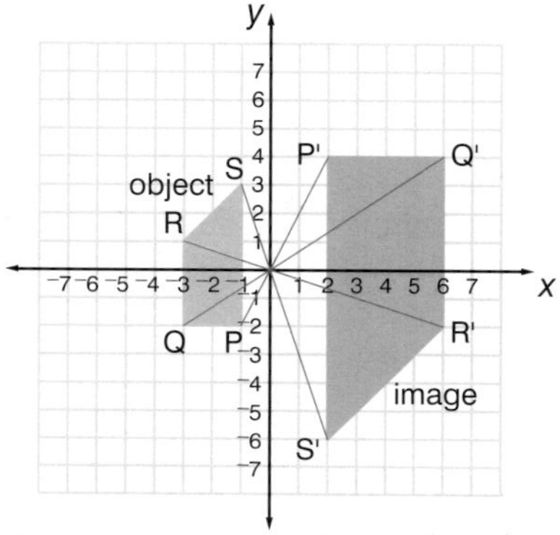

The points P, Q, R and S are drawn and the enlargement is produced to give P'Q'R'S'.

e.g. The triangle LMN with vertices L(2, 2), M(5, 2) and N(2, 8) is enlarged with a scale factor of $^-\frac{1}{3}$ about the point (0, 2). Draw LMN and hence L'M'N'.

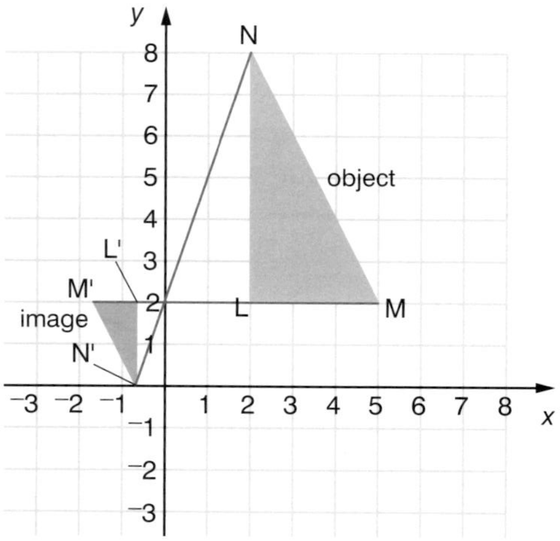

The points L', M' and N' and ΔL'M'N' are drawn and the enlargement produced to give L'M'N' as shown.

Shape, space & measures 20–27 (higher tier)

1 Solve $\cos\theta \leqslant {}^{-}0.5$ in the range $0 \leqslant \theta \leqslant 360°$.
(2 *marks*)

2 The volume of a cone of height 12.5 cm is 220 cm^3. What is the volume of a similar cone of height 5 cm?
(2 *marks*)

3 The arc length between two points A and B on the circumference of a circle is 10 cm. If the radius of the circle is 14.2 cm, calculate the angle subtended at the centre of the circle and the area of the sector.
(2 *marks for each part*)

4 The sides of a triangle are 2 cm, 3 cm and 4 cm respectively. Find the size of the largest angle and the area of the triangle.
(2 *marks for each part*)

5 Given that PT is a tangent to the circle at P and ∠PSQ = 23°, find ∠PRQ, ∠POQ and ∠QPT.

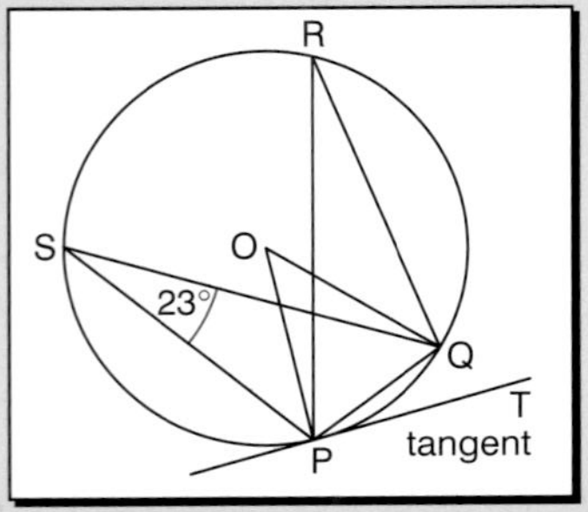

(1 *mark for each part*)

6 The diagram shows the triangle PQR where L, M and N are the midpoints of the sides as shown. Vector **PL** = **a** and vector **PN** = **b**.

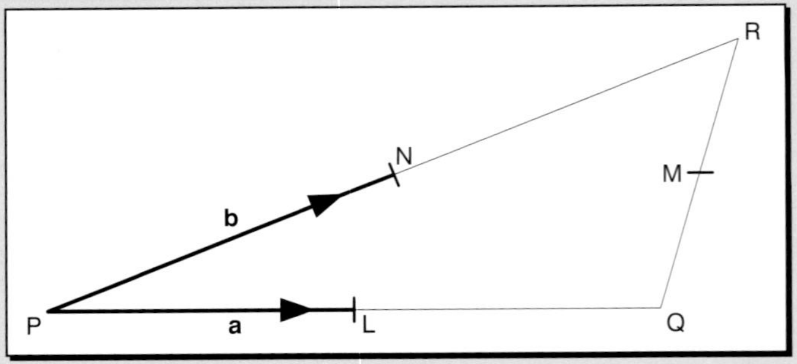

Express in terms of **a** and **b** the vectors:

(a) LN **(b) QR** **(c) PM**.

What can you say about LN and QR?
(1 *mark for each part and* 2 *marks for full explanation*)

***Total marks* = 20**

SCORE

1 $120° \leqslant \theta \leqslant 240°$

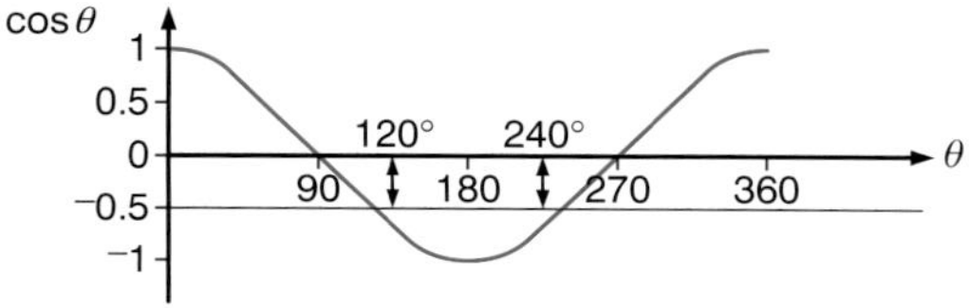

From the graph θ lies in the range $120° \leqslant \theta \leqslant 240°$.

2 $14.1\ cm^3$ (3 s.f.)

Ratio of lengths is $12.5 : 5 = 5 : 2$.

Ratio of volumes is $5^3 : 2^3 = 125 : 8$

$= 1 : \frac{8}{125}$

$= 220 : 14.08$

3 Angle = 40.3° (3 s.f.), area = $71\ cm^2$

$$\text{Arc length} = \frac{\theta}{360} \times 2\pi r$$

$$10 = \frac{\theta}{360} \times 2 \times \pi \times 14.2$$

$$\theta = 40.349\,141°$$

$$\text{Area} = \frac{40.349\,141}{360} \times \pi r^2$$

$$= 71\ cm^2$$

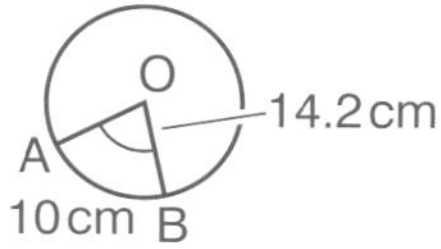

SCORE

4 104° (3 s.f.), $2.90\ cm^2$ (3 s.f.)

Largest angle is opposite longest side.

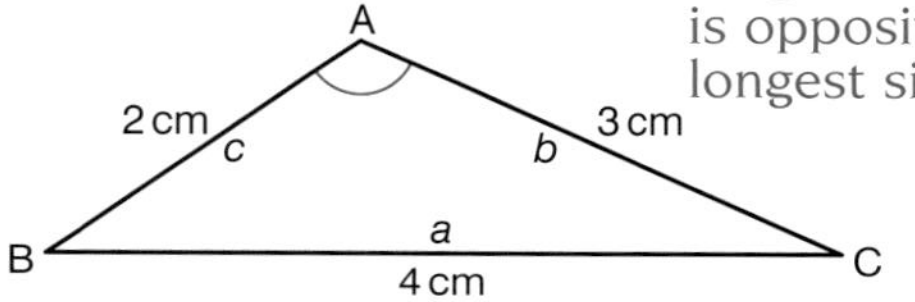

$$\text{Cosine rule: } \cos A = \frac{b^2 + c^2 - a^2}{2bc}$$

$$= \frac{9 + 4 - 16}{12} = -\frac{3}{12}$$

$\cos A = -\frac{3}{12} = -0.25$ so A = 104.477 51°

Area of triangle $= \frac{1}{2} bc\sin A$

$= \frac{1}{2} \times 2 \times 3 \times \sin 104.477\,51°$

$= 2.904\,737\,5\ cm^2$

5 ∠PRQ = 23° Angles subtended by the same arc.

∠POQ = 46° Angle at centre is twice angle at circumference.

∠QPT = 23° Alternate segment theorem.

6 **(a)** **b** – **a**   **LN** = –**a** + **b** = **b** – **a**

(b) 2(**b** – **a**)   **QR** = –2**a** + 2**b** = 2(**b** – **a**)

(c) **a** + **b**   **PM** = **PQ** + **QM** = **PQ** + ½**QR**

= 2**a** + ½ × 2(**b** – **a**)

= 2**a** + **b** – **a** = **a** + **b**

LN is parallel to QR and QR = 2LN or LN = ½QR.

TOTAL SCORE OUT OF 20

Representing data

You should always ensure that the representation chosen is appropriate to the data being represented.

Pictograms

A pictogram (or pictograph or ideograph) is a simple way of representing data. The frequency is indicated by a number of identical pictures. When using a pictogram, you must remember to include a key to explain what the individual pictures represent as well as giving the diagram an overall title.

e.g. Show the following information as a pictogram.

Ice cream sales

Flavour	vanilla	strawberry	raspberry	other
Sales	8	4	5	3

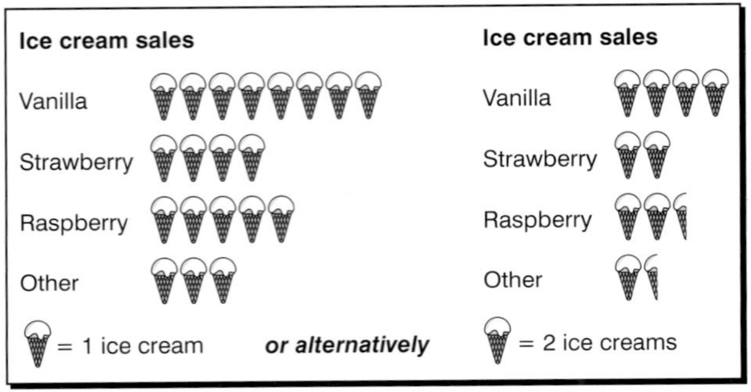

Bar charts

A bar chart is a common way of representing data. The frequencies are indicated by vertical or horizontal bars, all of the same width. When using a bar chart, you should clearly label the axes and give the diagram a title to explain what it represents.

e.g. Show the following information as a bar chart.

Drinks served

Drink	tea	coffee	milkshake	other
Sales	7	3	4	2

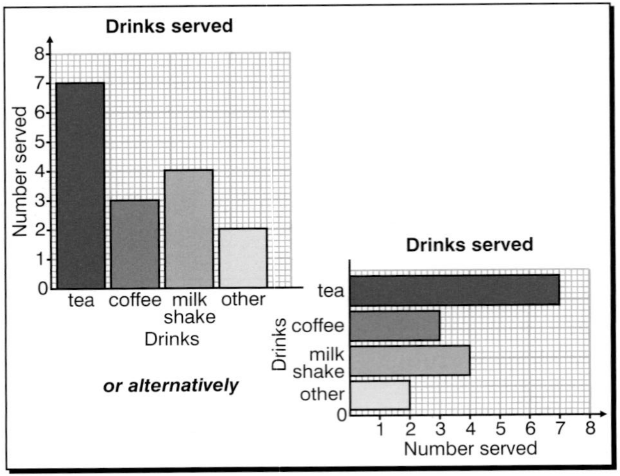

Line graphs

A line graph is another way of representing data. The frequencies are plotted at various points and joined by a series of straight lines. Once again, you must clearly label the axes and give the diagram a title to explain what it represents.

e.g. Show the following information as a line graph.

A patient's temperature

Time	07.00	08.00	09.00	10.00	11.00	12.00
Temperature (°F)	102.8	101.5	100.2	99.0	98.8	98.6

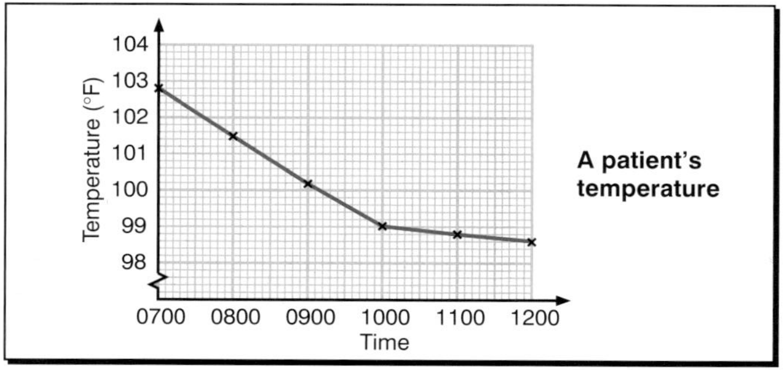

Pie charts

A pie chart is another common way of representing data. The frequency is represented by the angles (or areas) of the sectors of a circle. When using a pie chart, you must clearly label each of the sectors and give the diagram a title to explain what it represents.

e.g. Show the following information as a pie chart.

Travelling to work

Travel	bus	walk	cycle	car
Number	36	20	11	23

As there are 90 people, the pie chart needs to be drawn to represent 90 people. There are 360° in a full circle so each person will be shown by $\frac{360°}{90} = 4°$ of the pie chart.

Travel	Number	Angle
Bus	36	$36 \times 4° = 144°$
Walk	20	$20 \times 4° = 80°$
Cycle	11	$11 \times 4° = 44°$
Car	23	$23 \times 4° = 92°$
	Total = 90	Total = 360°

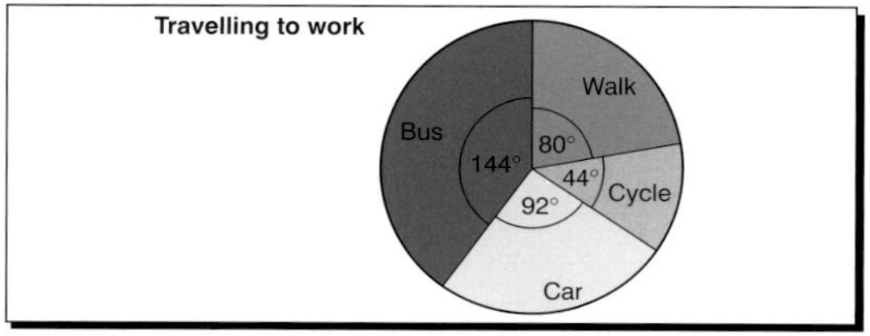

Stem and leaf diagrams

A stem and leaf diagram is a useful way to show information. It consists of a stem and a leaf which together form the data to be illustrated. When using stem and leaf diagrams, you must remember to explain the stem (with a key) and give your diagram a title, to describe what it represents.

e.g. The daily numbers of guests in a hotel over a two-week period were recorded.

36	42	42	51	46	32	31
29	27	34	41	43	37	22

Show this information on a stem and leaf diagram. Use the tens digit as the stem.

Number of guests

	2	9	7	2			
	3	6	2	1	4	7	
stem (tens)	4	2	2	6	1	3	leaf (units)
	5	1					Key: 4|2 means 42

The stem and leaf may also be shown as an ordered stem and leaf diagram, as follows.

Number of guests

	2	2	7	9			
	3	1	2	4	6	7	
stem (tens)	4	1	2	2	3	6	leaf (units)
	5	1					Key: 4|2 means 42

You can show two sets of data on one display if you use a back-to-back stem and leaf diagram.

e.g. The data shows the number of words in sentences for two newspapers.

Show the data as a stem and leaf diagram.

Express
11 23 34 28 15 35
36 38 22 36 17 27

Recorder
41 23 33 26 31 18
34 36 26 40 37 14

Express		*Recorder*
7 5 1	1	4 8
8 7 3 2	2	3 6 6
8 6 6 5 4	3	1 3 4 6 7
	4	0 1
Leaf (units)	↑ Stem (tens)	Leaf (units)

Measures of central tendency

Measures of central tendency are often referred to as measures of average i.e. **mode**, **median** and **mean**.

Mode

The mode of a distribution is the value that occurs most frequently.

e.g. Find the mode of the following distribution.

8, 6, 7, 4, 9, 8, 8, 6, 7, 6, 8

The number 8 occurs most frequently so the mode is 8.

NB The mode of a grouped frequency distribution has no meaning although you might be asked to find the modal group.

Median

The median of a distribution is the middle value when the values are arranged in numerical order.

e.g. Find the median of the following distribution.

8, 6, 7, 4, 9, 8, 8, 6, 7, 6, 8

Rearranging in numerical order:

4, 6, 6, 6, 7, 7, 8, 8, 8, 8, 9

So the median = 7

Where there are two middle values (i.e. when the number of values is even) then you add the two middle numbers and divide by 2.

e.g. Find the median of the following distribution.

8, 6, 7, 4, 9, 8, 8, 6, 7, 6, 8, 10

Rearranging in order:

4, 6, 6, 6, 7, 7, 8, 8, 8, 8, 9, 10

$$\text{So the median} = \frac{7 + 8}{2} = 7\tfrac{1}{2}$$

In general, the median is the $\frac{1}{2}(n + 1)$th value in the distribution, where n is the number of values in the distribution.

In the above example there are 12 values and the median is the $\frac{1}{2}(12 + 1)$th $= 6\frac{1}{2}$th value (indicating that it lies between the 6th and 7th values).

NB The median of a frequency distribution or a grouped frequency distribution can be found from the **cumulative frequency diagram** (see Handling data card 6).

Mean

The mean (or arithmetic mean) of a distribution is found by summing the values of the distribution and dividing by the number of values.

e.g. Find the mean of the following distribution.

8, 6, 7, 4, 9, 8, 8, 6, 7, 6, 8

Mean

$$= \frac{8 + 6 + 7 + 4 + 9 + 8 + 8 + 6 + 7 + 6 + 8}{11}$$

$$= \frac{77}{11} = 7$$

Mean of a frequency distribution

The mean of a frequency distribution is found by summing the values of the distribution and dividing by the number of values.

$$\text{Mean} = \frac{\text{sum of values}}{\text{number of value}} \text{ or}$$

$$\text{Mean} = \frac{\Sigma\text{frequency} \times \text{values}}{\Sigma\text{frequencies}} = \frac{\Sigma fx}{\Sigma f}$$

where Σ means 'the sum of' and Σf is equal to the number of values.

e.g. Find the mean of the following frequency distribution.

Value	5	6	7	8
Frequency	1	3	2	4

Value x	Frequency f	Frequency × value fx
5	1	$1 \times 5 = 5$
6	3	$3 \times 6 = 18$
7	2	$2 \times 7 = 14$
8	4	$4 \times 8 = 32$
	$\Sigma f = 10$	$\Sigma fx = 69$

$$\text{Mean} = \frac{\Sigma\text{frequency} \times \text{values}}{\Sigma\text{frequencies}} = \frac{\Sigma fx}{\Sigma f} = \frac{69}{10} = 6.9$$

Mean of a grouped frequency distribution

For a grouped frequency distribution you use the mid-interval values (or midpoints) as an 'estimate' of the interval.

e.g. Find the mean of the following grouped frequency distribution.

Group	10–20	20–25	25–30	30–50
Frequency	7	9	6	3

The midpoints of the individual groups are 15, 22.5, 27.5 and 40 respectively, from the table above.

Group	Frequency	Midpoint	Frequency × midpoint
10–20	7	15	105
20–25	9	22.5	202.5
25–30	6	27.5	165
30–50	3	40	120
	$\Sigma f = 25$		$\Sigma fx = 592.5$

$$\text{Mean} = \frac{\Sigma\text{frequency} \times \text{values}}{\Sigma\text{frequency}}$$

$$= \frac{\Sigma fx}{\Sigma f}$$

$$= \frac{592.5}{25}$$ Sum of the frequencies is 25.

$$= 23.7$$

Measures of spread

Measures of spread are another useful tool for comparing data. They tell how spread out, or consistent, the data are.

Range

The **range** of a distribution is found as the numerical difference between the greatest value and least value. The range should always be given as a *single* value.

e.g. Find the range of the following test marks.

9 7 8 10 9 8 8 2 9 10 8

Rearranging in order:

2 7 8 8 8 8 9 9 9 10 10

Greatest value = 10

Least value = 2

Range = greatest value – least value

= 10 – 2

= 8

NB The range of 8 is deceptive here as it is affected by the value of 2 (which is called an **extreme value** because it is not typical of the rest of the distribution).

Interquartile range

While the range can be affected by extreme values, the **interquartile range** only takes the middle 50% of the distribution into account.

The interquartile range is found by dividing the data into four parts and working out the difference between the **upper quartile** and the **lower quartile**.

e.g. Find the interquartile range of the following test marks.

9 7 8 10 9 8 8 2 9 10 8

Arrange the data in order and consider the middle 50% of the distribution.

2	7	8	8	8	8	9	9	9	10	10
		↑ **LQ**			↑ **median**			↑ **UQ**		

Lower quartile = 8

Upper quartile = 9

Interquartile range

= upper quartile – lower quartile

= 9 – 8

= 1

NB In general, the lower quartile is the $\frac{1}{4}(n + 1)$th value and the upper quartile is the $\frac{3}{4}(n + 1)$th value in the distribution.

Box and whisker plots

Another useful way of showing a frequency distribution is a box and whisker plot (sometimes called a box plot). The plot shows:

- the median
- the upper and lower quartiles
- the maximum and minimum values.

e.g. The insurance premiums paid by eleven households are listed below.

£340 £355 £400 £320 £380 £320
£632 £365 £340 £380 £370

Draw a box and whisker plot and calculate the interquartile range.

Arrange the data in order.

£320 £320 £340 £340 £355 £365 £370 £380 £380 £400 £632

↑ LQ (£340) ↑ Median (£365) ↑ UQ (£380)

Upper quartile = £380
Lower quartile = £340
Interquartile range = upper quartile – lower quartile
= £380 – £340
= £40

The box plot looks like this.

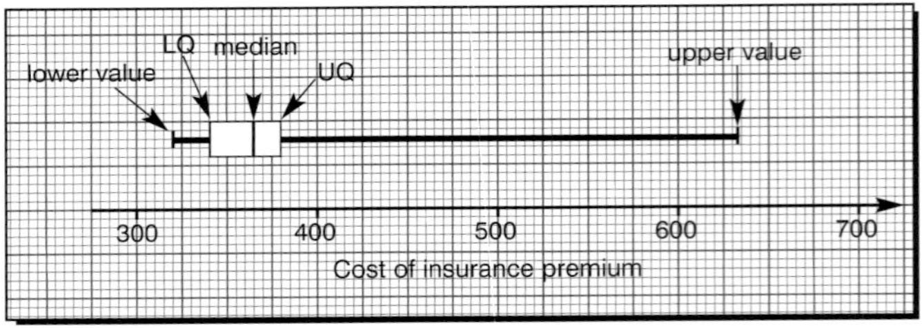

A box and whisker plot can be used to compare different distributions. It will enable you to comment on the spread of the data and make comparisons about the medians and quartiles.

Cumulative frequency diagrams

A cumulative frequency diagram (or **ogive**) can be used to find the median and quartiles of a distribution.

To find the cumulative frequency, find the accumulated totals and plot them against the data values. The cumulative frequency diagram is formed by joining the points with a smooth curve.

e.g. The following information shows the time (given to the nearest minute) which customers have to wait to be served.

Waiting time (minutes)	Frequency
1–3	6
4–6	11
7–9	20
10–12	13
13–15	5

Plot this information on a cumulative frequency diagram.

Waiting time (minutes)	Frequency	Cumulative frequency
1–3	6	6
4–6	11	17
7–9	20	37
10–12	13	50
13–15	5	55

NB The final cumulative frequency should equal the sum of the frequencies.

The cumulative frequency diagram looks like this.

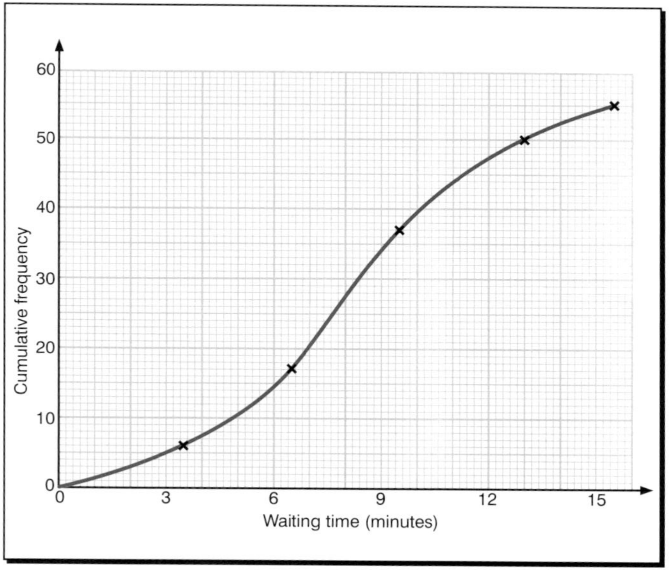

NB The cumulative frequencies must always be plotted at the **upper class boundaries** (i.e. 3.5, 6.5, 9.5, 12.5 and 15.5).

Remember that the interval of 1–3 minutes, to the nearest minute, includes all values from 0.5 minutes to 3.5 minutes, and so on.

Using the cumulative frequency (1)

The cumulative frequency diagram can be used to find the median and quartiles of the given distribution.

Median

You have already seen that the median is the middle value and, in general, the median is the $\frac{1}{2}(n + 1)$th value in the distribution (see Handling data card 4).

For the cumulative frequency diagram on Handling data card 8, the median is given by the $\frac{1}{2}(n + 1)$th value $= \frac{1}{2}(55 + 1) = 28$th value in the distribution.

Interquartile range

The interquartile range is the difference between the upper quartile and the lower quartile where the lower quartile is the $\frac{1}{4}(n + 1)$th value and the upper quartile is the $\frac{3}{4}(n + 1)$th value in the distribution (see Handling data card 6).

For the cumulative frequency diagram the lower quartile is given by the $\frac{1}{4}(n + 1) = \frac{1}{4}(55 + 1) = 14$th value and the upper quartile is given by the $\frac{3}{4}(n + 1) = \frac{3}{4}(55 + 1) = 42$nd value.

NB Your answers should usually be correct to the nearest half-square (in the example on the right, ± 0.1 minutes).

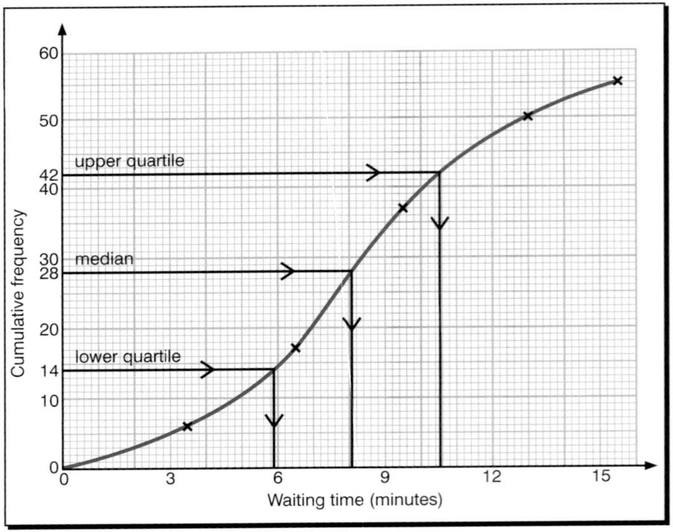

From the graph:

median $= 8.1$

upper quartile $= 10.6$

lower quartile $= 5.9$

interquartile range

$= \text{upper quartile} - \text{lower quartile} = 10.6 - 5.9$

$= 4.7$

Using the cumulative frequency (2)

The cumulative frequency diagram can also be used to find out other information about the distribution such as how many customers waited:

(a) less than 5 minutes

(b) more than 10 minutes.

Waiting time (minutes)	Frequency	Cumulative frequency
1–3	6	6
4–6	11	17
7–9	20	37
10–12	13	50
13–15	5	55

(a) To find out how many customers waited less than 5 minutes, read the cumulative frequency for a waiting time of 5 minutes. From the graph, the number is 10.5. An answer rounded to 10 or 11 would be acceptable.

(b) To find out how many customers waited more than 10 minutes, read the cumulative frequency for a waiting time of 10 minutes.

From the graph, the number of customers who waited less than 10 minutes is 39.5.

So the number of customers who waited more than 10 minutes is 55 – 39.5 = 15.5, or 16 customers to the nearest whole number.

NB This information cannot just be read from the table as the 10–12 group contains times from 9.5 minutes to 12.5 minutes.

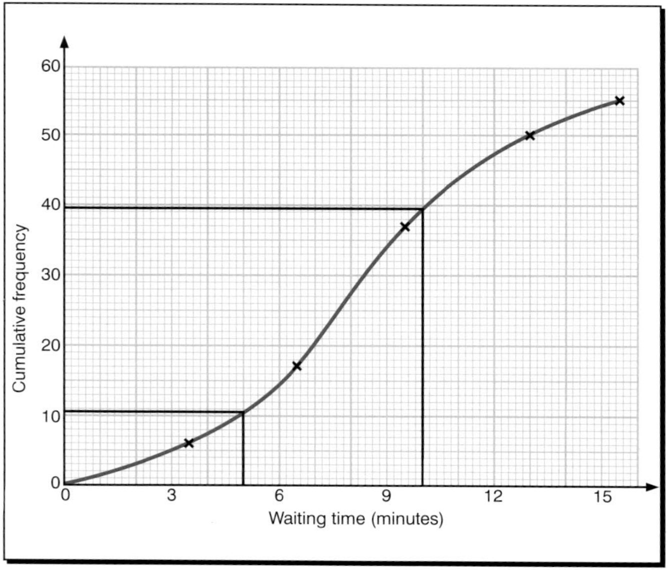

Handling data 1–10

1 The following distribution shows the number of different flavoured ice creams sold one morning.

Flavour	Frequency
Vanilla	15
Strawberry	8
Raspberry	3
Other	4

Show this information as:

(a) a pictogram
(b) a bar chart
(c) a pie chart.

(3 marks)

2 Study this pie chart.

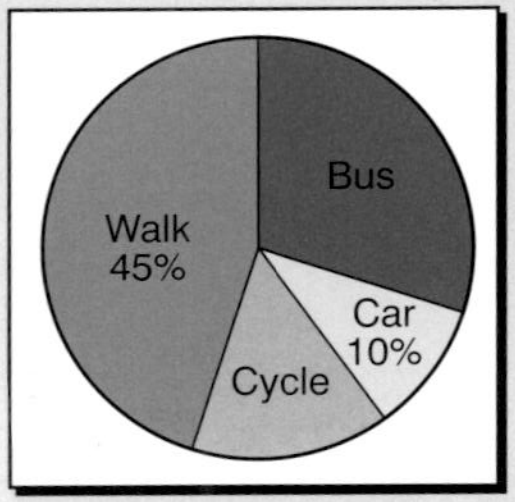

(a) Which method of transport was the most popular?
(b) What method of transport was the least popular?
(c) What angle is represented by the 'car' sector?

The number of students who travelled by bus is twice the number who cycled. 100 students who took part in the survey.

(d) How many students travelled by bus?

(4 marks)

3 Find the mean, median and mode of the following frequency distribution.

Shoe size	5	$5\frac{1}{2}$	6	$6\frac{1}{2}$	7
Frequency	3	4	10	13	4

(4 marks)

4 Calculate an estimate of the mean for the following data.

Height (inches)	10–20	20–30	30–40	40–50
Frequency	13	19	10	6

(2 marks)

5 The box plots show the length of sentences in two books.

(a) Find the range for the first book. *(1 mark)*

(b) Find the interquartile range of the second book. *(1 mark)*

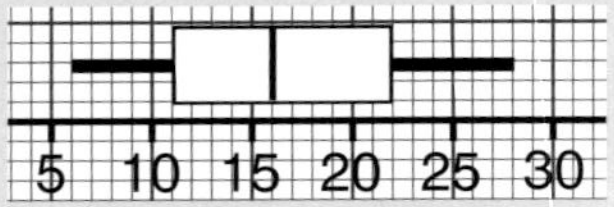

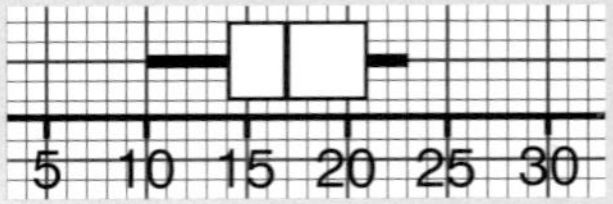

6 The following table shows the number of words per paragraph in a newspaper. Draw a cumulative frequency diagram to illustrate this information and use your graph to estimate:

Words	Number of paragraphs
1–10	15
11–20	31
21–30	45
31–40	25
41–50	4

(a) the median and interquartile range
(b) the percentage of paragraphs over 35 words in length.

(5 marks)

Total marks = 20

102

1 (a)

(b)

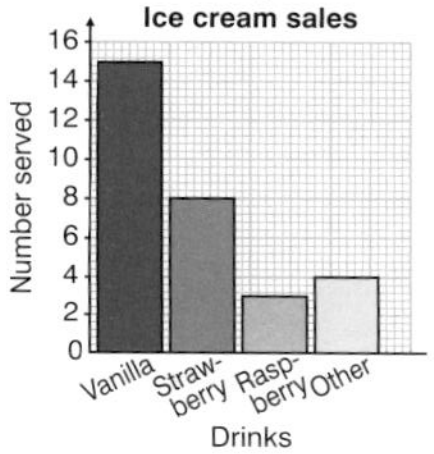

(c)

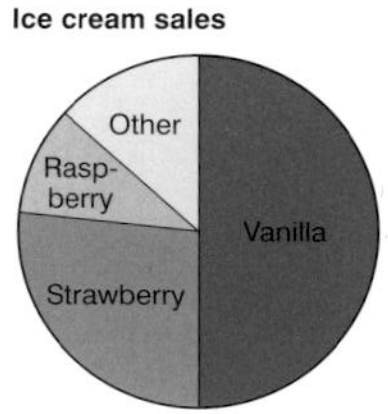

2 (a) Walking — From the pie chart.
(b) Car — From the pie chart.
(c) 36° — 10% of 360° = 36°
(d) 30 — The number travelling by bus or cycling = 45% of 100 = 45 people and the number travelling by bus is 30.

3 Mean = 6.16 (3 s. f.) $\Sigma fx = 209.5$ and $\Sigma f = 34$

$$\text{mean} = \frac{\Sigma fx}{\Sigma f} = 6.16 \text{ (3 s. f.)}.$$

Median = $6\frac{1}{2}$ $\frac{1}{2}(35 + 1) = 18$th value which occurs at $6\frac{1}{2}$.

Mode = $6\frac{1}{2}$

4 27 inches to an appropriate degree of accuracy. Using the mid-interval values of 15, 25, 35 and 45, $\Sigma fx = 1290$ and $\Sigma f = 48$

$$\text{mean} = \frac{\Sigma fx}{\Sigma f} = 26.875$$

5 (a) 22 **(b)** 7

6

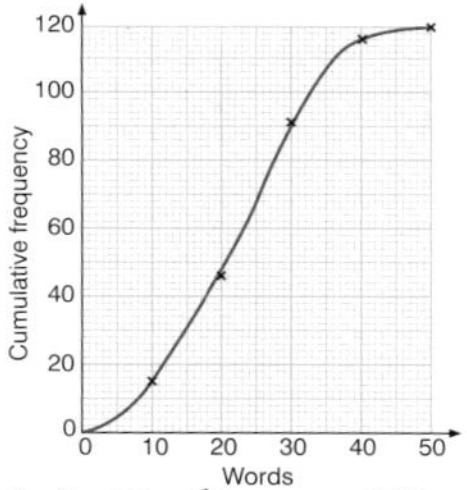

The cumulative frequencies should be plotted at the upper boundaries of 10, 20, 30, 40, etc. as the number of words is discrete in this instance.

(a) Median = 23, upper quartile = 29, lower quartile = 15 All found by drawing suitable lines on the graph.
Interquartile range
= upper quartile – lower quartile
= 29 – 15 = 14

(b) 10% Number of paragraphs under 35 words = 108.
Number of paragraphs over 35 words
= 120 – 108 = 12

$$\text{Percentage of paragraphs} = \frac{12}{120} \times 100 = 10\%$$

SCORE

SCORE

TOTAL SCORE OUT OF 20

Scatter diagrams

Scatter diagrams (or **scatter graphs**) are used to show the relationship between two variables. Each of the two variables is assigned to a different axis and the information is plotted as a series of coordinates on the scatter diagram.

e.g. Draw a scatter graph of the following table which shows the heights and shoe sizes of 10 pupils.

Shoe size	3	2	$6\frac{1}{2}$	4	3	6	1	$3\frac{1}{2}$	5
Height (cm)	133	126	158	135	128	152	118	142	150

To draw the scatter graph, consider each pair of values as a different pair of coordinates (3, 133), (2, 126), ($6\frac{1}{2}$, 158), etc.

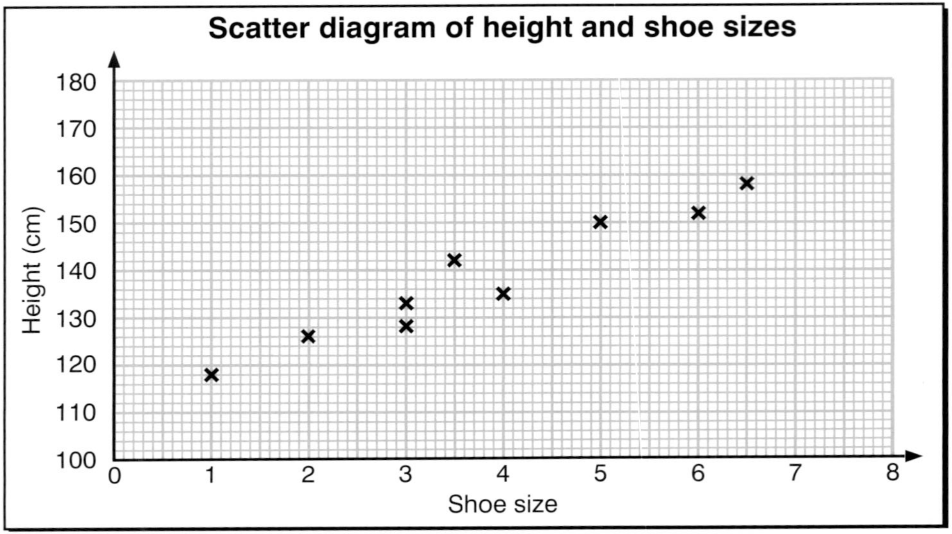

Correlation

Scatter graphs can be used to show whether there is any relationship or **correlation** between the two variables. The following descriptions of such relationships or correlation need to be known.

Strong correlation

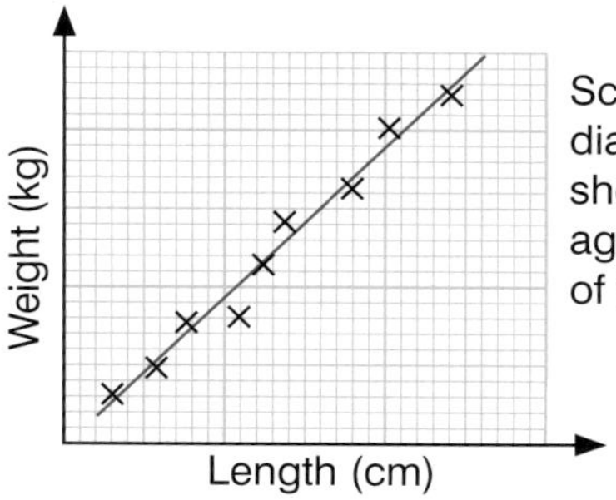

Scatter diagram to show length against weight of fish.

Moderate correlation

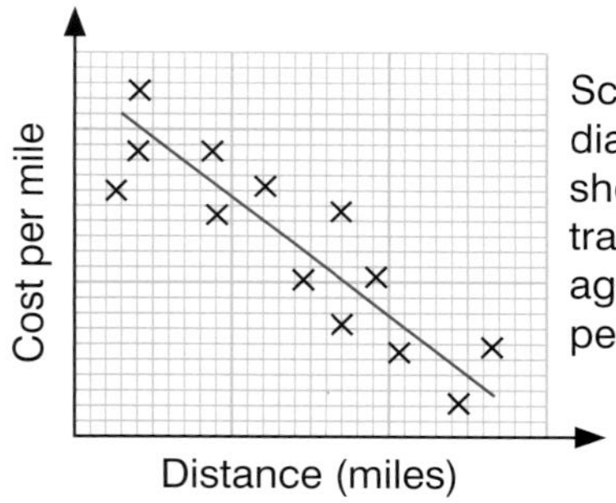

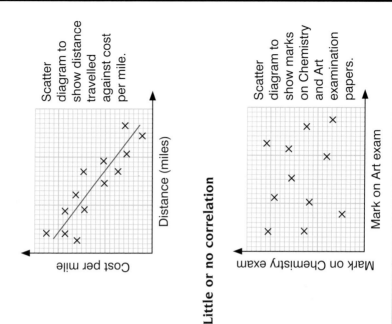

Scatter diagram to show distance travelled against cost per mile.

Little or no correlation

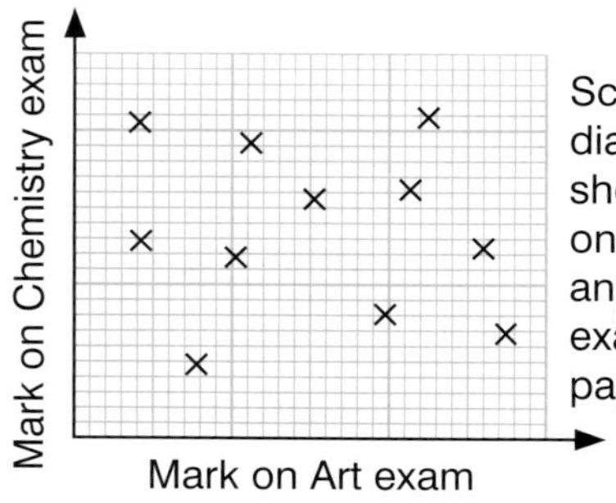

Scatter diagram to show marks on Chemistry and Art examination papers.

Positive and negative correlation

Correlation can also be described in terms of positive and negative correlation.

Positive correlation

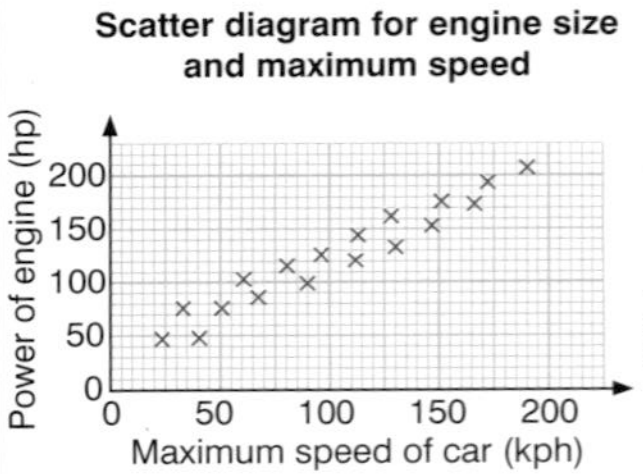

Negative correlation

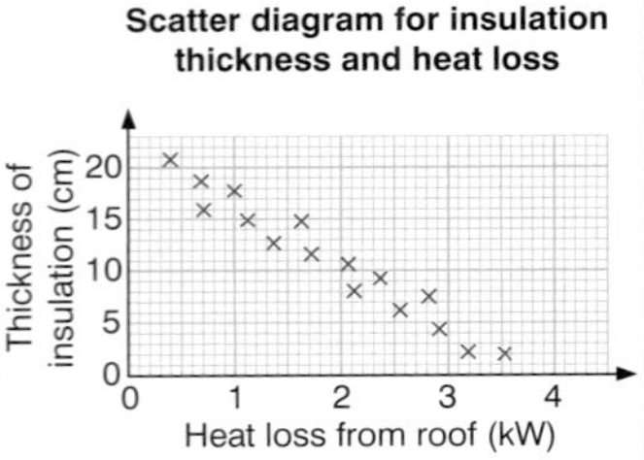

Where an increase in one variable is associated with an increase in the other variable, the correlation is **positive** or **direct**; where an increase in one variable is associated with a decrease in the other variable, the correlation is **negative** or **inverse**.

Line of best fit

When the points on a scatter diagram show moderate or strong correlation, a straight line can be drawn through, or as close to, as many of the points as possible, to approximate the relationship. This is the **line of best fit** (or **regression line**). It can be used to predict other values from the given data.

e.g. Draw a scatter graph and line of best fit for the following table which shows the heights and shoe sizes of 10 pupils.

Shoe size	3	2	$6\frac{1}{2}$	4	3	6	1	$3\frac{1}{2}$	5
Height (cm)	133	126	158	135	128	152	118	142	150

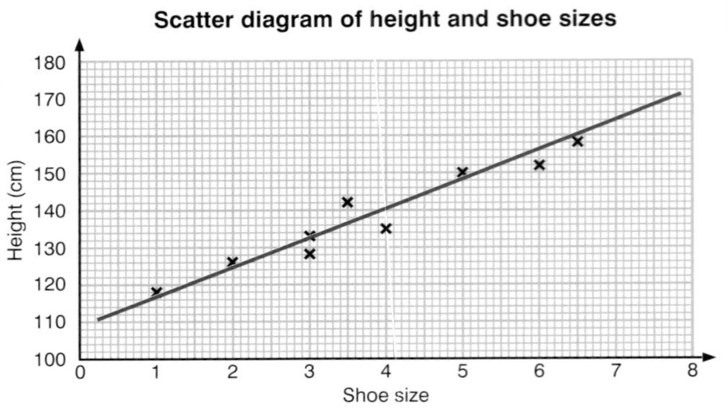

NB In most cases a line of best fit can be drawn 'by eye'. For more accurate work the line of best fit should pass through $(\overline{x},\overline{y})$ where $\overline{x}$ and $\overline{y}$ are the mean values of x and y respectively.

Time series (1)

Data that changes or fluctuates over time is best shown using a time series graph.

An example of this is shown in the following data about heating costs.

Year 1	Winter	£55
	Spring	£40
	Summer	£8
	Autumn	£25
Year 2	Winter	£63
	Spring	£48
	Summer	£16

As you can see, the costs change according to the season but they do not follow a clear pattern or rule, so they cannot be shown on a simple linear graph.

This information can be plotted on the following graph.

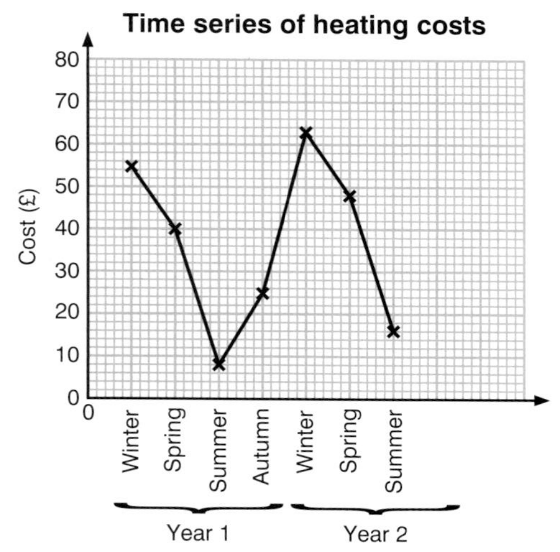

Time series (2)

Seasonal trends can be averaged out using moving averages. In this case a four-point moving average is appropriate and can be found by averaging successive four points at a time.

The first four-point moving average

$$= \frac{£55 + £40 + £8 + £25}{4} = £32$$

The second four-point moving average

$$= \frac{£40 + £8 + £25 + £63}{4} = £34$$

The third four-point moving average

$$= \frac{£8 + £25 + £63 + £48}{4} = £36$$

The forth four-point moving average

$$= \frac{£25 + £63 + £48 + £16}{4} = £38$$

The four-point moving average can be plotted on the graph as shown.

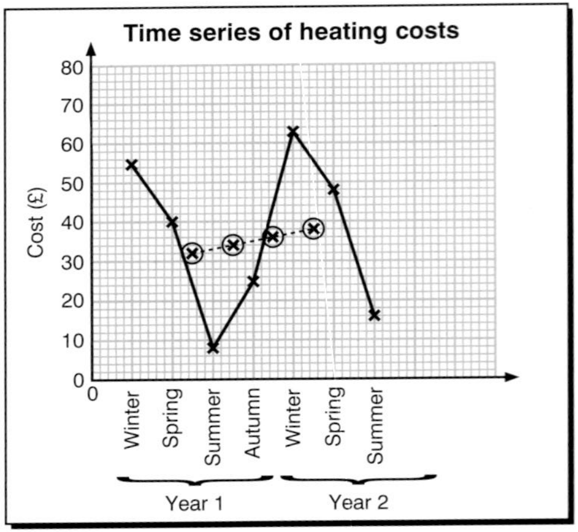

Note that the first four-point moving average is plotted in the 'middle' of the first four points etc.

The moving average is useful to identify trends and from the graph you can see that the trend is upwards.

Probability

In probability, an event which is certain to happen will have a probability of 1 whereas an event which is impossible will have a probability of 0. In general,

$$p(\text{success}) = \frac{\text{number of 'successful' outcomes}}{\text{number of possible outcomes}}$$

where p(success) means 'the probability of success'.

e.g. If a box contains seven red counters, ten blue counters and eight yellow counters then:

- the probability of picking a red counter is:

$$p(\text{red counter}) = \frac{\text{number of red counters}}{\text{number of counters}} = \frac{7}{25}$$

- the probability of picking a blue counter is:

$$p(\text{blue counter}) = \frac{\text{number of blue counters}}{\text{number of counters}} = \frac{\cancel{10}^{2}}{\cancel{25}_{5}} = \frac{2}{5}$$

Cancelling down where possible.

- the probability of picking a green counter is:

$$p(\text{green counter}) = \frac{\text{number of green counters}}{\text{number of counters}} = \frac{0}{25} = 0$$

i.e. the outcome is impossible.

Total probability

The probability of an event happening is equal to 1 *minus* the probability of the event not happening.

e.g. If the probability that it will rain tomorrow is $\frac{1}{5}$ then the probability that it will not rain tomorrow is $1 - \frac{1}{5} = \frac{4}{5}$.
i.e. $p(\text{not rain}) = 1 - p(\text{rain}) = 1 - \frac{1}{5} = \frac{4}{5}$

Theoretical and experimental

Theoretical probability is based on equally likely outcomes and is used to predict how an event should perform in theory; **experimental probability** (or **relative frequency**) is used to predict how an event performs in an experiment.

e.g. The following frequency distribution is obtained when a die is thrown 100 times.

Score	1	2	3	4	5	6
Frequency	15	19	18	15	17	16

The theoretical probability of a score of six is $\frac{1}{6}$ and the experimental probability (or relative frequency) of a score of six is $\frac{16}{100} = \frac{4}{25}$.

NB The expected number of sixes when a die is thrown 100 times is:

$$100 \times \frac{1}{6} = 16.\dot{6}$$

or 17 (to the nearest whole number).

Possibility spaces

A possibility space is a diagram which can be used to show the outcomes of various events.

e.g. If two fair dice are thrown and the sum of the scores is noted, then the following diagram can be drawn to illustrate the possible outcomes.

		Second die					
		1	2	3	4	5	6
First die	1	2	3	4	5	6	7
	2	3	4	5	6	7	8
	3	4	5	6	7	8	9
	4	5	6	7	8	9	10
	5	6	7	8	9	10	11
	6	7	8	9	10	11	12

The diagram above is a possibility space; there are 36 possible outcomes.

From the diagram:

- the probability of a sum of 2 is $\frac{1}{36}$ (as there is only one entry with a sum of 2)
- the probability of a sum of 7 is $\frac{6}{36} = \frac{1}{6}$ (as there are six entries with a sum of 7)
- the probability of a sum of 10 is $\frac{3}{36} = \frac{1}{12}$ (as there are three entries with a sum of 10)

etc.

Tree diagrams

A tree diagram is a diagram where the probabilities of different events are written on different branches, so that the probabilities on individual pairs (or groups) of branches always add up to one.

e.g. A bag contains five blue and four green discs. A disc is drawn from the bag, replaced and then a second disc is drawn from the bag. Draw a tree diagram to show the various possibilities that can occur.

The following tree diagram shows the various possibilities that can occur.

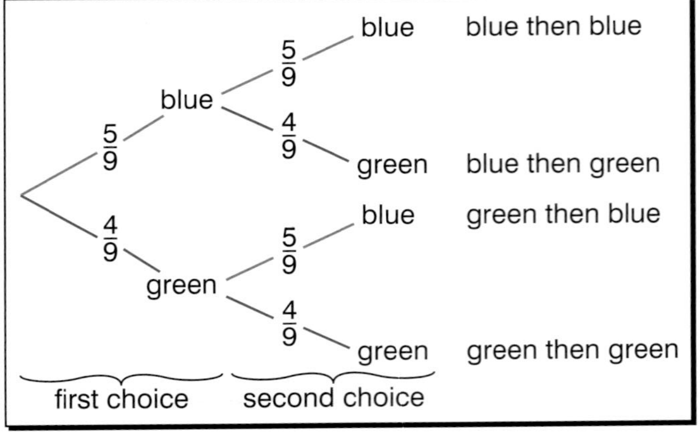

Mutually exclusive events

Events are **mutually exclusive** if they cannot both (or all) happen at the same time.

In the case of mutually exclusive events, apply the **addition rule** (also called the **or rule**):

p(A **or** B) = p(A) + p(B)

e.g. A spinner has 10 sides numbered 1 to 10.

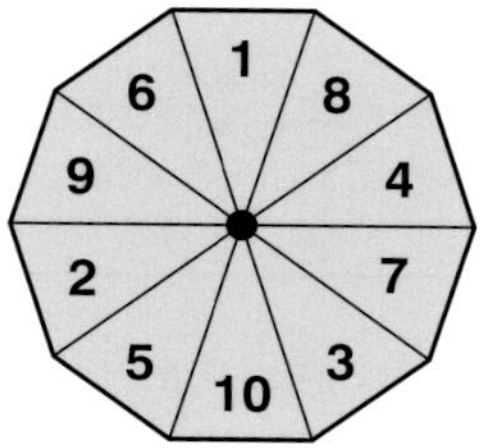

- The probability of scoring a five = p(5) = $\frac{1}{10}$
- The probability of scoring a five or a six
  = p(5 or 6)
  = p(5) + p(6) As the events are mutually exclusive.
  = $\frac{1}{10} + \frac{1}{10} = \frac{2}{10} = \frac{1}{5}$ Cancelling down.

- The probability of scoring a multiple of 3 or a multiple of 4
  = p(multiple of 3 or multiple of 4)
  = p(multiple of 3) + p(multiple of 4) As the events are mutually exclusive.
  = p(3 or 6 or 9) + p(4 or 8)
  = p(3) + p(6) + p(9) + p(4) + p(8) As the events are mutually exclusive.
  = $\frac{1}{10} + \frac{1}{10} + \frac{1}{10} + \frac{1}{10} + \frac{1}{10} = \frac{5}{10} = \frac{1}{2}$

- The probability of scoring a multiple of 2 or a multiple of 3
  = p(multiple of 2 or multiple of 3)

  These events are **not** mutually exclusive as the number 6 is common to both events and if the probabilities are added then p(6) will be added twice.

  So p(multiple of 2 or multiple of 3)
  = p(2 or 3 or 4 or 6 or 8 or 9 or 10)
  = p(2) + p(3) + p(4) + p(6) + p(8) + p(9) + p(10) All of these events are now mutually exclusive.
  = $\frac{1}{10} + \frac{1}{10} + \frac{1}{10} + \frac{1}{10} + \frac{1}{10} + \frac{1}{10} + \frac{1}{10}$
  = $\frac{7}{10}$

Independent events

Events are independent if one event can occur without affecting the other events.

For independent events, use the **multiplication rule** (also called the **and rule**):

p(*A* **and** *B*) = p(*A*) × p(*B*)

e.g. A bag contains four red counters and three blue counters. A counter is drawn from the bag, replaced and then a second counter is drawn from the bag. Draw a tree diagram to illustrate this situation.

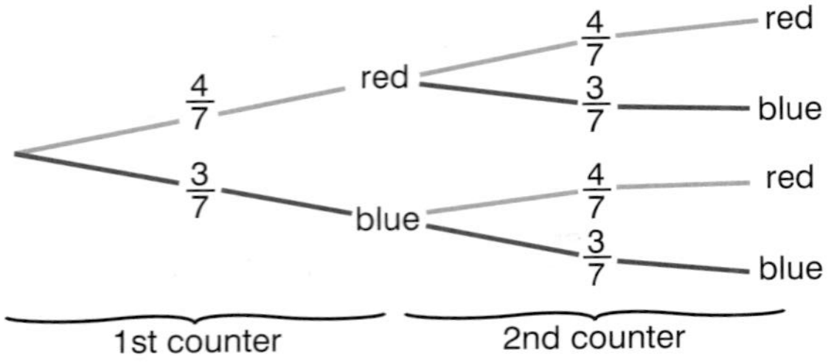

- The probability that both counters will be red
  = p(red and red)
  = p(red) × p(red)
  As the events are independent.
  $= \frac{4}{7} \times \frac{4}{7} = \frac{16}{49}$

- The probability that both counters will be blue
  = p(blue and blue)
  = p(blue) × p(blue)
  As the events are independent.
  $= \frac{3}{7} \times \frac{3}{7} = \frac{9}{49}$

- The probability that the first counter will be red and the second counter will be blue
  = p(red and blue)
  = p(red) × p(blue)
  As the events are independent.
  $= \frac{4}{7} \times \frac{3}{7} = \frac{12}{49}$

- The probability that one counter will be red and one counter will be blue is not the same as the last example.
  No order is specified, so you must find the probability that the first counter will be red and the second counter will be blue **or** the first counter will be blue and the second counter will be red.
  = p(red and blue **or** blue and red)
  = p(red and blue) + p(blue and red)
  As the events are mutually exclusive.
  = p(red) × p(blue) + p(blue) × p(red)
  As the events are independent.
  $= \frac{4}{7} \times \frac{3}{7} + \frac{3}{7} \times \frac{4}{7}$
  $= \frac{12}{49} + \frac{12}{49}$
  $= \frac{24}{49}$

Handling data 11–19

1 The following information shows the marks awarded to students on two examination papers.

German	30	39	25	10	45	15	53	39	32	50
French	30	56	40	14	64	28	79	48	44	59

Plot these points and comment on your findings.

Draw the line of best fit and use it to calculate the other mark of someone who gets:
(a) 65 in French **(b)** 65 in German.
Which answer is the more accurate? Why?
(4 marks)

2 A manufacturer produces components. In a sample of 500 it is found that 12 are faulty.
(a) Find the probability that a component is faulty.
(b) How many faulty components would you expect in a batch of 100?
(2 marks)

3 The probability that a plane arrives early is 0.08 and the probability that it arrives on time is 0.63. What is the probability that it arrives late?
(1 mark)

4 In an experiment, a coin is tossed and a die is thrown. Draw a possibility space and use it to calculate the probability of scoring:
(a) a head and a six
(b) a tail and an odd number.
(3 marks)

5 One bag holds three red, four blue and two orange counters. Another holds five yellow and two green counters. Draw a labelled tree diagram to show the possibilities when a counter is picked from each bag. Calculate the probability of picking out:
(a) a red and a green counter
(b) a blue and a yellow counter. *(3 marks)*

6 The table shows the attendance rates in a school.

Date	Aut 98	Spr 99	Sum 99	Aut 99	Spr 00	Sum 00
% attendance	88	83	90	89	85	95

(a) Plot the information on a graph.
(b) Calculate the first three-point moving average.
The remaining three-point moving averages are 87.3, 88, 89.7.
(c) Plot all the values on the moving average graph.
(d) Comment on what your graph shows. *(4 marks)*

7 A bag holds red, blue or green cubes numbered 1, 2 or 3. These ase the probabilities for these cubes.

		Colour of cube		
		Red	Blue	Yellow
Number	1	0.2	0	0.1
	2	0.2	0.1	0.1
	3	0.1	0.1	0.1

Find the probability that a cube taken at random is:
(a) red and numbered 3 **(b)** blue or numbered 1
(c) red or numbered 2. *(3 marks)*

Total marks = 20

SCORE

1 Scatter diagram for marks in French and German

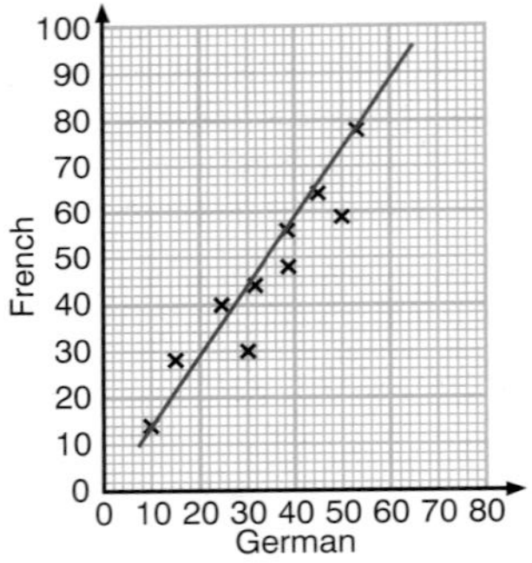

The points show strong a positive correlation.
(a) 46 marks Reading from the graph.
(b) 96 marks Reading from the graph.
The first answer is more accurate as it lies within the given papers' range of marks.

2 **(a)** $\frac{3}{125}$ $\frac{12}{500} = \frac{3}{125}$
(b) 2 (to an appropriate degree of accuracy)
$100 \times \frac{3}{125} = 2.4$.

3 0.29 p(late) = 1 – p(early or on time)

4

		Score on die					
		1	2	3	4	5	6
Coin	Head	H1	H2	H3	H4	H5	H6
	Tail	T1	T2	T3	T4	T5	T6

(a) $\frac{1}{12}$ From the table.
(b) $\frac{1}{4}$ From the table ($\frac{3}{12} = \frac{1}{4}$).

SCORE

5

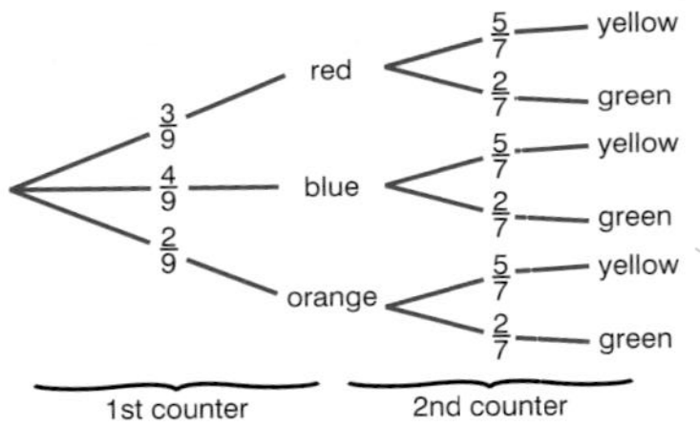

(a) $\frac{2}{21}$ p(red and green counter) $= \frac{3}{9} \times \frac{2}{7} = \frac{6}{63}$
$= \frac{2}{21}$.
(b) $\frac{20}{63}$ p(blue and yellow counter) $= \frac{4}{9} \times \frac{5}{7} = \frac{20}{63}$.

6 **(a)**

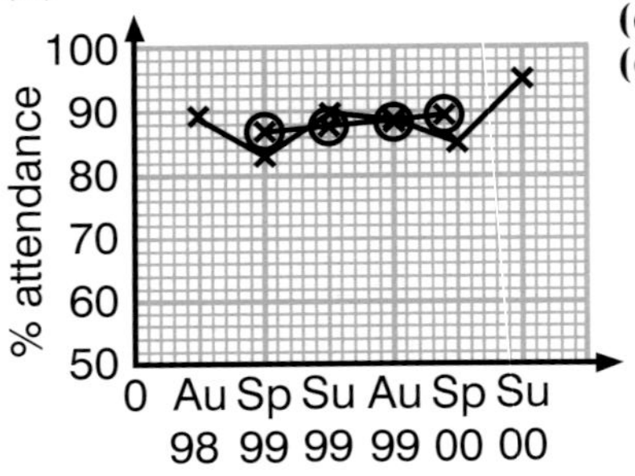

(b) 87
(c) see **(a)**
(d) Increasing

7 **(a)** 0.1
(b) 0.5 0.1 + 0.1 + 0.2 + 0.1 = 0.5
(c) 0.7 0.2 + 0.2 = 0.1 + 0.1 + 0.1 = 0.7
NB Do not count 'red and numbered two' twice.

TOTAL SCORE OUT OF 20

Histogram – unequal class intervals

Histograms are like bar charts except that it is the area of each 'bar' that represents the frequency.

The bars of the histogram should be drawn at the class boundaries and the area of the bars should be proportional to the frequency,

i.e. class width × height = frequency so that
height = frequency ÷ class width.

The height is referred to as the frequency density so:
frequency density = frequency ÷ class width

e.g. The following frequency distribution shows the height of 38 shrubs measured to the nearest centimetre. Draw the histogram.

Height (cm)	5–9	10–14	15–19	20–29	30–50
Frequency	4	8	10	9	7

To draw a histogram you need to calculate the frequency densities as follows.

Height	Frequency	Class width	Frequency density
5–9	4	5	4 ÷ 5 = 0.8
10–14	8	5	8 ÷ 5 = 1.6
15–19	10	5	10 ÷ 5 = 2.0
20–29	9	10	9 ÷ 10 = 0.9
30–50	7	21	7 ÷ 21= 0.3333...

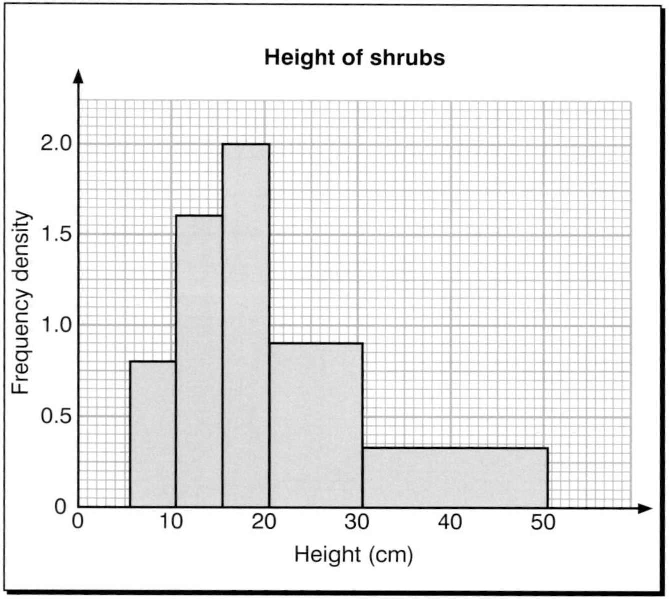

NB Since the height is measured to the nearest centimetre then the 5–9 interval extends from 4.5 to 9.5, giving it a width of 5 cm, etc.

Frequency polygons

A frequency polygon can be drawn from a bar chart or histogram by joining the midpoints of the tops of the bars, in order, with straight lines to form a polygon.

The lines should be extended to the horizontal axis so that the area under the frequency polygon is the same as the area under the bar chart or histogram.

e.g. Draw the histogram for the following information.

Height (cm)	5–9	10–14	15–19	20–29	30–50
Frequency	4	8	10	9	7

To draw the frequency polygon, calculate the frequency densities as before.

Height	Frequency	Class width	Frequency density
5–9	4	5	$4 \div 5 = 0.8$
10–14	8	5	$8 \div 5 = 1.6$
15–19	10	5	$10 \div 5 = 2.0$
20–29	9	10	$9 \div 10 = 0.9$
30–50	7	21	$7 \div 21 = 0.3333...$

The frequency polygon is then plotted at the mid-interval values i.e. 7, 12, 17, 24.5 and 40.

NB It is often helpful to draw the frequency polygon over the top of the completed histogram.

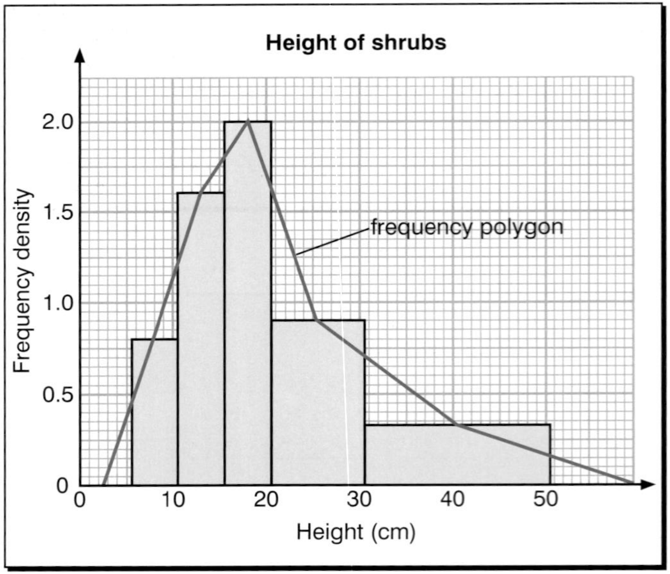

NB The lines are extended to the horizontal axis on each side so that the area under the frequency polygon is the same as the area under the histogram (see page 108).

Dependent events

Two or more events are **dependent** if one event happening affects the other event happening as shown in the following example.

e.g. A bag contains four red and three blue counters. A counter is drawn from the bag and then a second counter is drawn from the bag.

With replacement (independent events)
Where the first counter is replaced before the second counter is drawn then the two events are **independent** and the probabilities for each event are the same.

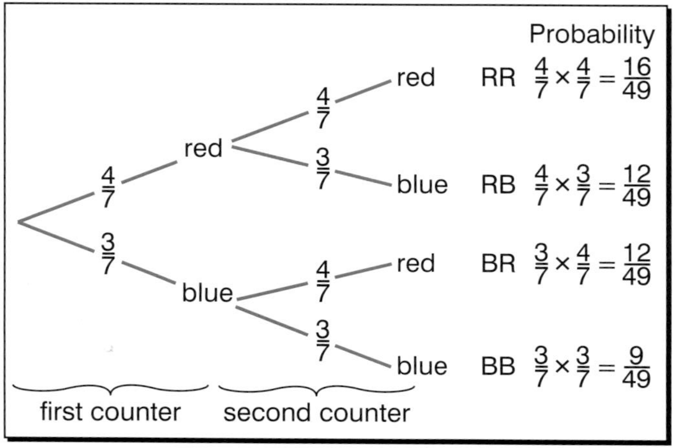

Without replacement (dependent events)
Where the first counter is not replaced before the second counter is drawn then the two events are not independent (i.e. they are dependent) and the probabilities on the second event will be affected by the outcomes on the first event.

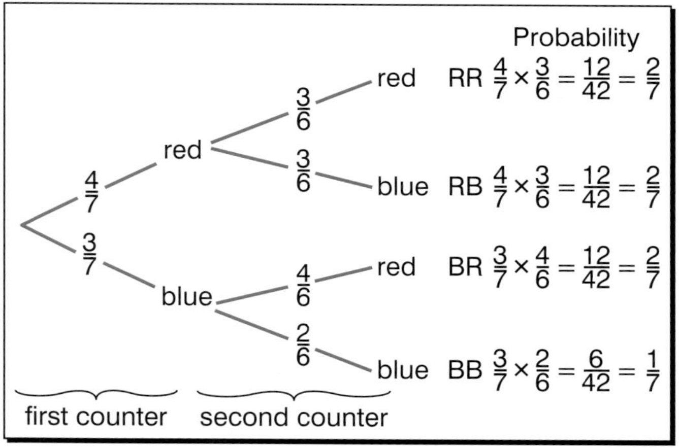

NB In the second diagram, the probabilities for the second counter depend upon the outcome of the first counter.

Sampling methods

Sampling should be representative of the population as a whole and sample sizes should be as large as possible. You will need to know the following sampling techniques.

Convenience sampling

In convenience sampling or opportunity sampling, the first people who come along are chosen.

Convenience sampling might involve sampling friends and members of your own family and is therefore likely to involve some element of bias.

Random sampling

In random sampling, each member of the population has an equally likely chance of being selected.

Random sampling might involve giving each member of the population a number and then choosing the numbers at random, using some appropriate means to generate random numbers.

Systematic sampling

Systematic sampling uses random sampling with some system involved in choosing the member of the population to be sampled.

Systematic sampling might include numbering each member of the population according to some system (e.g. by their name, age, height, etc.).

Stratified sampling

In stratified sampling, the population is divided into groups or strata. From each stratum you choose a random or systematic sample so that the sample size is proportional to the size of the group in the population as a whole.

e.g. In a class where there are twice as many girls as boys, the sample should include twice as many girls as boys.

Quota sampling

Quota sampling involves choosing population members who have specific characteristics which are decided beforehand.

Quota sampling is popular in market research where interviewers are told that:

- there must be equal numbers of men and women or
- there must be twice as many adults as teenagers or
- the sample should include ten shoppers and five commuters

etc.

Handling data 20–23

1 The populations of three villages are given in the table.

Village	Population
Lasminster	2500
Marton	4100
Newcliffe	5900

What is the most appropriate way to take a sample of 240 villagers?

(2 marks)

2 The table shows the weights of 50 parcels.

Weight (kg)	Frequency
6–10	4
11–15	14
16–25	21
26–35	8
36–55	3

Draw a histogram to represent these data.

(3 marks)

3 Draw a frequency polygon for the date shown in question 2.

(3 marks)

4 A bag contains four red and three blue balls. One ball is chosen at random and its colour is noted. The ball is not replaced, then a second ball is chosen at random and its colour noted. Draw a tree diagram to represent this situation and use it to calculate the probability of obtaining:

(a) two red balls

(b) one ball of each colour.

(3 marks for diagram and 2 marks for each part)

5 The probability that Ben passes the driving theory test on his first attempt is $\frac{7}{9}$. If he fails then the probability that he passes on the next attempt is $\frac{6}{7}$. Draw a tree diagram to represent this situation and use it to calculate the probability that Ben passes the driving test on his third attempt.

(3 marks for diagram and 2 marks for probability)

Total marks = 20

SCORE

1 Stratified sampling.

Population of all villages = 12 500.

Lasminster: $\frac{2500}{12\,500} \times 240 = 48$

Marton: $\frac{4100}{12\,500} \times 240 = 78.72 = 79$

Newcliffe: $\frac{5900}{12\,500} \times 240 = 113.28 = 113$

All answers given to the nearest integer.

2

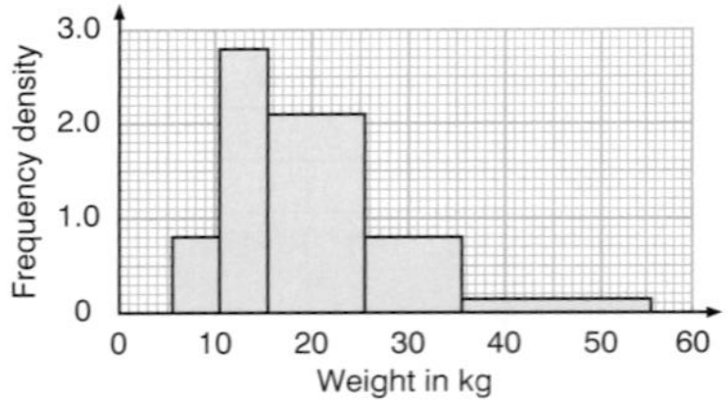

NB For the weights 6–10, the boundaries are 5.5 and 10.5 so the class width = 5, etc.

Weight (kg)	Frequency	Width	Frequency density
6–10	4	5	$4 \div 5 = 0.8$
11–15	14	5	$14 \div 5 = 2.8$
16–25	21	10	$21 \div 10 = 2.1$
26–35	8	10	$8 \div 10 = 0.8$
36–55	3	20	$3 \div 20 = 0.15$

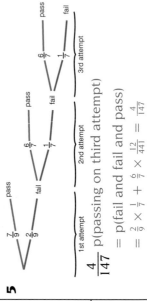

SCORE

3

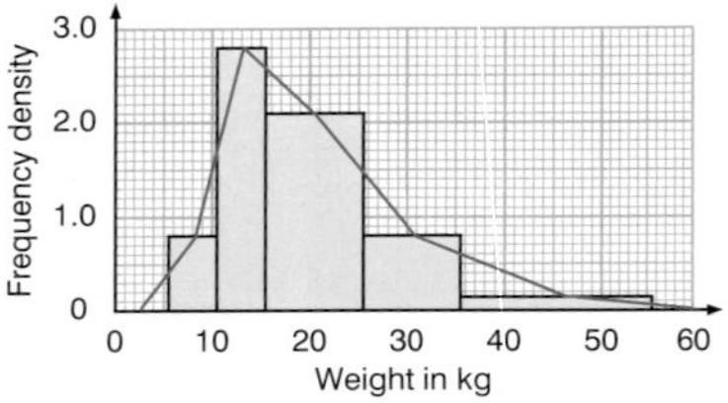

4

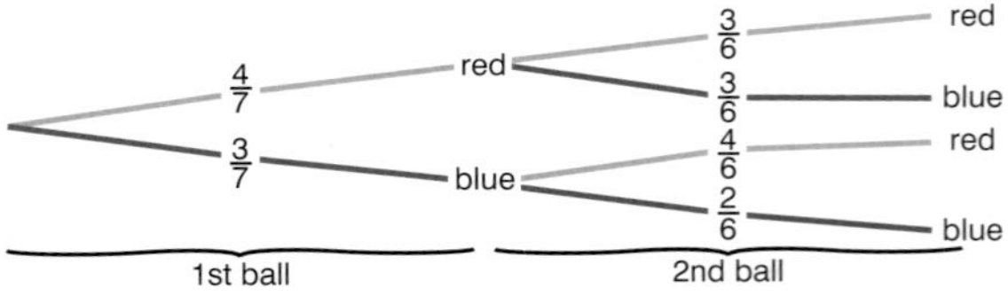

(a) $\frac{2}{7}$ $\quad \frac{4}{7} \times \frac{3}{6} = \frac{2}{7}$

(b) $\frac{4}{7}$ $\quad \frac{4}{7} \times \frac{3}{6} + \frac{3}{7} \times \frac{4}{6} = \frac{4}{7}$

5

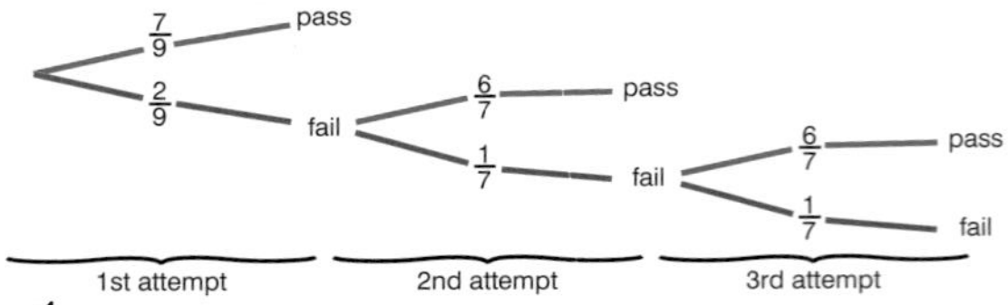

$\frac{4}{147}$ p(passing on third attempt)

= p(fail and fail and pass)

$= \frac{2}{9} \times \frac{1}{7} + \frac{6}{7} \times \frac{12}{441} = \frac{4}{147}$

TOTAL SCORE OUT OF 20

INTERMEDIATE TIER SCORE CHART (1)

Topic	*Check yourself*	Marks out of 20
Number 1–10	1	
Number 1–10	2	
Number 11–14	3	
Number 11–14	4	
Algebra 1–6	6	
Algebra 1–6	7	
Algebra 7–12	8	
Algebra 7–12	9	
Shape, Space & Measures 1–11	11	
Shape, Space & Measures 12–19	12	
Shape, Space & Measures 12–19	13	
Handling Data 1–10	15	
Handling Data 11–19	16	

Plot your marks for each *Check yourself* on the grid and then read across for your grade.

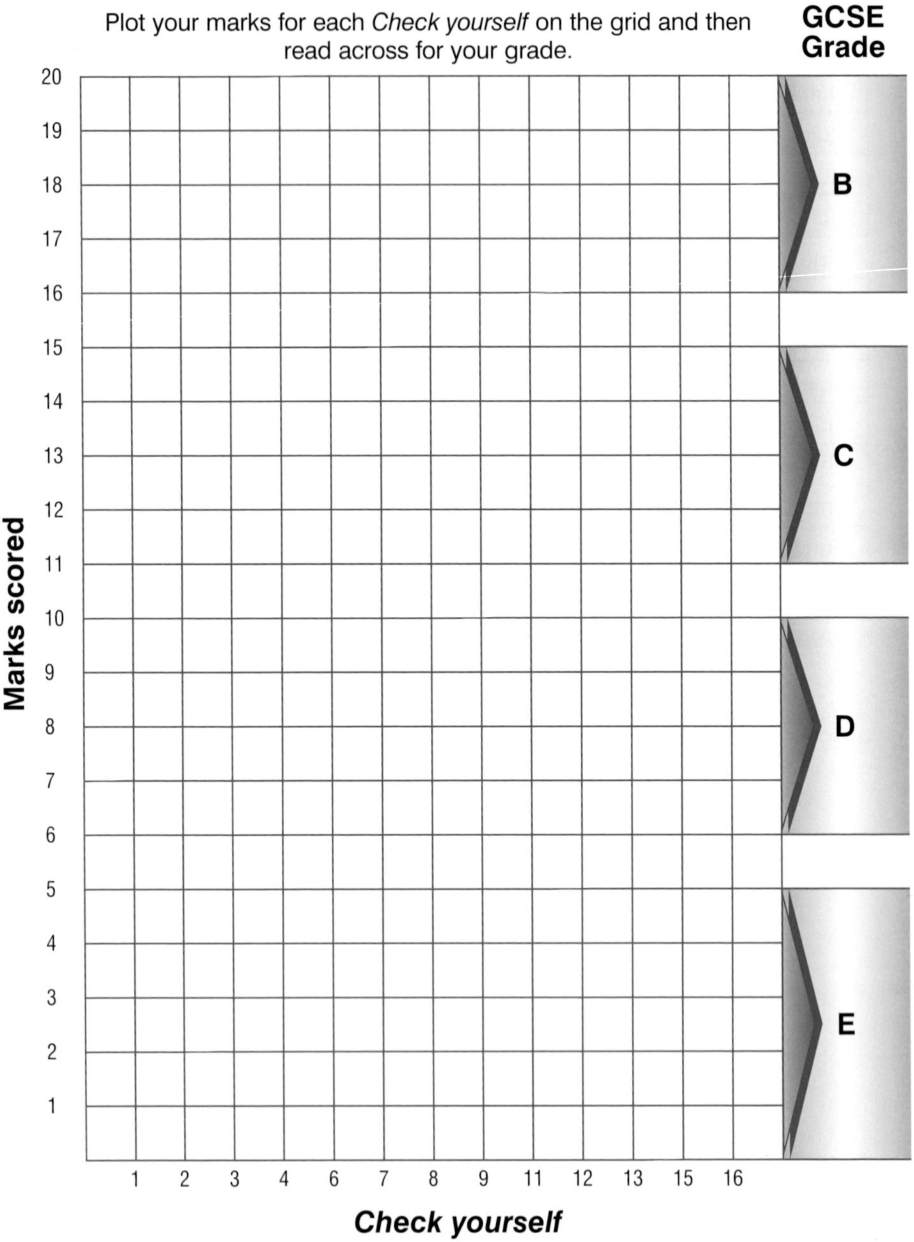

Higher Tier Score Chart (1)

Topic	*Check yourself*	Marks out of 20
Number 15–18	5	
Algebra 13–18	10	
Shape, Space & Measures 20–27	14	
Handling Data 20–23	17	

Plot your marks for each *Check yourself* on the grid and then read across for your grade.

GCSE Grade

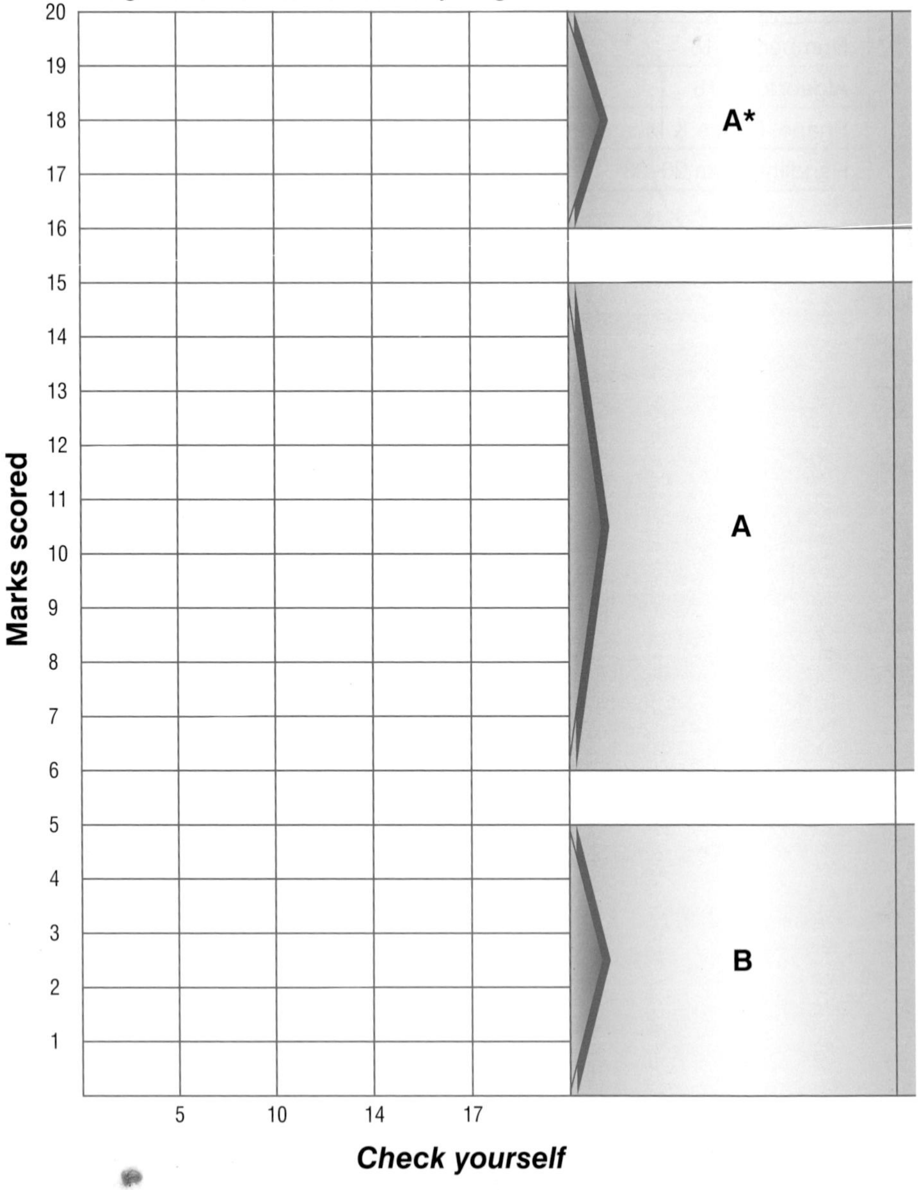